The Social Neuroscience of Intergroup Relations: Prejudice, can we cure it?

Image retrieved from: https://pixabay.com/ (All images and videos on Pixabay are released free of copyrights under Creative Commons CC0. You may download, modify, distribute, and use them royalty free for anything you like, even in commercial applications. Attribution is not required)

Sylvia Terbeck

The Social Neuroscience of Intergroup Relations: Prejudice, can we cure it?

 Springer

Sylvia Terbeck
School of Psychology
Plymouth University School of Psychology
Plymouth, UK

ISBN 978-3-319-46336-0   ISBN 978-3-319-46338-4   (eBook)
DOI 10.1007/978-3-319-46338-4

Library of Congress Control Number: 2016955542

This Springer imprint is published by Springer Nature
The registered company is Springer International Publishing AG
The registered company address is: Gewerbestrasse 11, 6330 Cham, Switzerland

This book is dedicated to Dr Laurence Paul Chesterman.

Foreword

In the *Oxford English Dictionary*, the word 'prejudice' is defined as 'preconceived opinion not based on reason or actual experience'. However, the Dictionary goes on to note that in more recent times, the notion of prejudice specifically depicts 'unreasoned dislike, hostility, or antagonism towards, or discrimination against, a race, sex, or other class of people'. Considering the staggering amount of global violence and suffering apparently based on ethnic, religious, and political differences, there could hardly be a more important topic but what is the role for neuroscience? How could a brain-based approach help us understand and deal with this overwhelming problem, which seems self-evidently a matter for political and sociocultural transformation?

Sylvia Terbeck's fascinating book does not side step this challenge but explains with great clarity and accessibility how the brain ultimately is the basis of all behaviour and that a scientific understanding of prejudice in no way neglects the political and philosophical dimensions of this critical problem. Indeed neuroscience has been able to make substantial progress in uncovering the implicit cognitive biases and emotional responses and underpinning neural circuitry that mechanistically drive the psychological processes involved in the intergroup behaviours – all too often expressed as prejudice of various kinds.

One of the most powerful aspects of science is its potential to predict and control aspects of the natural world, which in the context of neuroscience includes human behaviour. Could an appropriate drug therefore be helpful in combating prejudice? Here Sylvia's discussion is particularly illuminating and well-informed, based as it is on some fascinating studies that she personally conducted with a widely used drug called propranolol, which apparently has the remarkable ability to decrease a laboratory measure of implicit prejudice. However, Sylvia's reservations about the use of drugs to produce moral (or any other kind of) 'enhancement' are convincing and highly topical.

I recommend this book to anyone wishing to understand how modern neuroscience can be applied to the analysis of fundamental human behaviours, even those that have caused strife and misery throughout recorded history. It takes a very accomplished author to integrate in a readily comprehensible way, the neuroscience approach with sociological, philosophical and political insights and this is what Sylvia has achieved. Her book deserves to be widely read.

University of Oxford   Phil Cowen
Oxford, UK

Acknowledgements

Firstly, I would like to thank Dr Laurence Paul Chesterman for his invaluable help. I would like to thank Prof Phil Cowen for his excellent foreword, but also for being my supervisor at Oxford University; the work could have not been completed without him. I would also like to thank my first supervisor Prof Miles Hewstone, also Dr Guy Kahane, as well as Prof Julian Savulescu. I would have never thought about moral enhancement – the topic Prof Julian Savulescu developed – otherwise. Thank you also to Dr Sarah McTavish for your support and all members of the Oxford Centre for Magnetic Resonance Imaging.

Very valuable comments and editing were conducted by my best friend Dr Ann Dowker; thank you. Also thank you to my friend Uma Shahani, who made great comments and suggestions. Thank you to Dr Bill Simpson, for discussing the book with me.

Furthermore, many thanks to the 2015/2016 class of undergraduate social psychology students from Plymouth University for comments, language editing, and ideas, especially to Ella Dowden, Tom Harlow, Lillian Hawkins, Georgia Lewis, Hana Tomaskova, Catherine Senior, Shannon Jackson, David Bennett, Fatin Soufieh, Abbie Cunningham, Shauna Barratt, Molly Russell, Jessica Haigh, Dean Moreton, Nicole Keslake, Maria Presley, Amie Barlow, and Nicole Gayler.

Contents

Chapter 1
Introduction

He would get up at 6 am sharp. People these days seem to be getting up later, 7, or even 8. Do those people not know that the early bird catches the worm? Usually he would start his day with 20 min of sit ups and crunches, but not today. He could feel the slight pain in his head from drinking too much alcohol last night. Once you start drinking it's difficult to just have a single beer, but then you regret that you cannot do all your duties the next day. It was more of a feeling of still being tired and disoriented, but he knew it would pass. Stefanie was already up; she was always out of bed before him. He knew that she would make an effort to get up just those few minutes before him. Once she said to him that it was not appropriate for a man to see his wife in a state of non- perfection, and of course that would be the case with her in the morning. He liked that she did that; that she cared so much about looking perfect for him that she would get up those few minutes earlier. Besides, she would want to make breakfast, and lay the table to his satisfaction. The second thing that many men did not seem to realise was that breakfast was the most important meal of the day. You should have breakfast like a king, lunch like a queen, dine like a pauper. He did not like sweet things for breakfast, but she would still put a pot of home-made marmalade on the table, just in case someone might want it someday. He would always have a hardboiled egg that was cooked in hot water for 5 min. She would be a little nervous when he cracked it, as she knew he would show his special face, when it was not a perfect mixture of hard and soft. The yolk should be slightly runny, but not too much, or it would spill over the rim of the eggshell, which would look messy. Of course he would get dressed for breakfast, shave, have a shower, and brush his teeth. Admittedly, it does not taste nice to eat food when you still have leftover toothpaste on your tongue, but that cannot be helped. There was one thing that made him feel good about himself, his position, and his achievements every time he put on his clothes that his wife would have found for him and laid on the bedside stool the evening before; it was his shoes. What makes a true gentleman, a man of honour and respect, a man that has achieved what a man should achieve; hand-made shoes. Often men bought designer shoes; those imported from America. But he could recognise them immediately, and he would know that this man was

© Springer International Publishing Switzerland 2016
S. Terbeck, *The Social Neuroscience of Intergroup Relations: Prejudice, can we cure it?*, DOI 10.1007/978-3-319-46338-4_1

only an imposter, not someone who could truly afford hand-made shoes. It was 6.15 when he started his breakfast with an egg, toast, cold meat selection, orange juice and black coffee. He did not read the newspapers these days but would rather fully enjoy the food. Stefanie would not talk at breakfast time; it was a time of eating, not talking. "Have a nice day, and be careful." she would say at 6.30 when breakfast was over and he would take his coat to go to work.

Today they were mostly doing the counting, and selecting, and not the shooting. Anyway, it was good that he had been recently promoted and that he was thus mostly involved with office work and selection processes. Only the young ones would need to actively do the shooting these days, so it would be a lazy day with not much new to expect. At 7 work would start, and he would sit behind his desk with a long list of bastards, criminals, and other folks to sort out. It was good that they had recently developed the number system as it made the process much more effi-cient and speedy. Those allocated to the right would go to the working group and those allocated to the left would go to the gas chamber. Sorting folks was following rules; such as those who look strong should work. Strong is defined as hard shoul-ders, clean teeth, wide legs, white outer eyes, which is mostly easy to see. It is usu-ally pretty noisy, mostly females make loud noises. He would be wearing a loaded gun, which he would use to shoot them – in self-defence – if they attacked, if their noise levels rose, or if they were non-compliant and no other method seemed appro-priate. All morning would go as usual and he would select Jews for work or as redundant. He would usually look at his watch at 11am, which was just in time for a second breakfast. He would have hot black coffee, which his wife had prepared in his thermos flask. He would only have a small sandwich, which his wife would carefully wrap in aluminium foil. Of course he would have to bring the foil back home; what a waste to throw away the foil every day if it was only used once for a sandwich. Today she made him ham and salad, with a little butter, which was good as otherwise the butter would drip onto the aluminium foil and would make it dirty, so that it could not be used again.

"Ok, I am ready to continue." This was at 11.15 am, as the breakfast break was always shorter than the lunch break, which was 1 h. "You to the right", it was a female with strong qualities, "You to the left", it was a youngster, about 11 years of age. He deemed her to be redundant because he discovered some disease related issue; her eyes looked funny, there was clearly some problem with her ability to move her pupils accurately, which he had seen before, and he thought was some inherited problem. It makes you always look funny, he was thinking and you can never be sure if their vision is impaired. "That is my daughter, please. Sir, please, I will not go without my daughter, she is only a child. Please Sir." the woman was shouting and crying in despair. This was bad, because now the child was panicked by seeing her mother in such a state. They had not eaten for a long time. The child felt safe initially, felt as if she was quite enjoying the close work with her mother, but it was getting less and less; less and less food, less and less hope. She saw more and more people crying, more and more not understanding what was happening in this world. She could not help herself, now seeing her mother, in such a state as she broke down in tears. All these days of horror, when would it end? When could she play again? It was 11.21 am when he shot both of them. After all they were breaking the rules; being too loud, and would not follow the order any more, and most of all, why not? They were only Jews.

1.1   What Is Happening in the Brain of Such a Person?

How can someone become so cruel, so inhuman, so lacking in remorse and guilt, so 'prejudiced' that they can forget their basic human instinct to empathise with a little child? Could there be something, some neurons, or some networks in the brain that functioned abnormally? And if so, is he ill? And if so, can this be cured? And if so, should this be cured? This book will give an insight into these questions, and will explore the nature of human prejudice and of humans' tendency to be 'bad'. But before I give a further outline of the book, I first want to address the question: Is he just an isolated exceptional case of a bad person? Of course most readers might have been surprised, and maybe even a little shocked, when finding that this man, who is a bit pedantic about his breakfast eggs, who has a lovely caring wife, then goes to work to kill innocent people. Indeed, I could feel myself getting annoyed with the fictional character as I was describing his second breakfast and how he would find it a waste to throw away the aluminium foil even though he had just thrown away a person's life. You might now think; well, it is only a fictional character, one I created for this book. I did. But of course we all know that there were horrendous crimes during the 2nd World War, just like the ones I described. Concentration camps were real, innocent children being shot actually happened, and the people who committed these acts and then went home to their wives, leading a normal life. And of course, we also know that prejudice creates or contributes to war and murder all the time. In Joseph Stalin's communist regime, millions of people were killed. Recently, in January 2012, more than 3000 people were killed as two tribes in South Sudan went to war. The people killed were not only soldiers; they were also women and children. Indeed, entire villages were destroyed and burned. Killing someone because they are not of ones' own race, own religion, or own group happens all the time. Prejudice can fuel killing, torture and other forms of cruelty and make people forget that the one they are approaching is a person, just like them. Of course it is often more complicated than this. Nazi Germans, for example, who were involved in the killings in the concentration camps might of course not (only) have committed those crimes because they were prejudiced, but also because they had to follow orders or because many who did not obey were killed themselves. We know about very brave Germans, or indeed people all over the world, who hid Jews, and protested about Hitler's dictatorship.

Now people might think that this was in the past, or this is happening somewhere far away. There might be isolated terrible monsters that do this, and that it is not relevant to modern democratic societies. You might think that *normal* humans today are not really that prejudiced. People might think that people today are not capable of doing such terrible acts. Of course, now, in Europe, America, and in most western democracies, the majority of people do not kill someone because of prejudice. Even though hate crimes do of course occur, there are fewer in comparison to a war situation. In fact, all our current social laws would forbid acts of prejudice and discrimination. So what is the problem here? Well, there are two problems: One is that people now, in Europe, America, and democratic societies, *do* seem to have a bias of favouring their own over a different group; be it in terms of race, religion, age, or gender. And the second problem is that as the situation and social laws change, people can change as well. Below I will describe an experiment, which illustrates what normal people are capable of doing if the situation changes. In fact this experiment was conducted partly as a consequence of what was observed in the Second World War. Indeed, many people believed that those who were involved in the killings of innocent Jews were extreme sadists, not normal people. This experiment will demonstrate how cruelly normal people can behave. It was conducted over 40 years ago in 1971, at Stanford University. I think some readers might know what is coming now; I am going to describe one of the most notorious experiments in the study of human psychology; Prof. Phillip Zimbardo's Stanford prison experiment. Twenty one volunteers were screened and rated as healthy, not suffering from any mental disorders. They were divided into two groups: some were the prisoners, some were the guards. They were brought into the basement of the university which was made up as a fake prison, with fake cells, prison walls and so forth. The participants were wearing either prisoners' or prison guards' uniforms. Then Prof. Zimbardo observed. After just a few days, "S*uddenly, the whole dynamic changed as they believed they were dealing with dangerous prisoners, and at that point it was no longer an experiment*" said Prof. Zimbardo[1]. After only a few days the guards became very cruel, treated prisoners harshly, shouted at them and humiliated them. They seemed to have had completely forgotten that just about 48 h ago they were all the same. The role and the environment changed them so quickly. Guards made prisoners strip naked, put bags over their heads and made them complete harsh exercise. The experiment, which was to last 2 weeks, was terminated after 6 days, as a number of the "prisoners" broke down. Again, participants in this study were compiled of average people. *"The study is the classic demonstration of the power of situations and systems to overwhelm good intentions of participants and transform ordinary normal young men into sadistic guards or for those playing prisoners to have mental breakdowns."* Prof. Zimbardo commented.

Now I want to come back to the question I posed at the beginning: "Is he just an exceptionally bad man?" During the Second World War numerous factors contributed to people's actions; people *did* have to follow orders or were risked being killed themselves. Thus there may have been reasons other than prejudice or group membership that led people to perform these acts. Numerous researchers have in fact found many other contributing factors, for example obedience to authority (Atran 2003; Swanson 2015), collective identity, or a charismatic leader. Most likely many factors come together to make people behave in 'monstrous' ways. Indeed, it might be suggested that morality – what is regarded as good and bad – can

---

[1] Citations taken from http://www.bbc.co.uk/news/world-us-canada-14564182

be shaped or influenced by social and cultural norms, and maybe even by suggestions of what other people say. In the fictional book "The kind worth killing" (by Peter Swanson), Lily, who in the end kills everyone who she does not like, says to Ted (a husband upset about his wife's infidelity):" Truthfully, I don't think murder is necessarily as bad as people make it out to be. Everyone dies. What difference does it make if a few bad apples get pushed along a little sooner than God intended to? And your wife, for example, seems like the kind worth killing."

Coming back to the question of prejudice, it should however have become clear that: Prejudice is not a problem of the past and it's not only happening somewhere far away. Also it is wrong to assume that modern 'normal' people could never behave in a cruel manner; they can, and they often do. Two factors can combine and reinforce each other: firstly people do have a bias to favour their own over other groups (I shall discuss this in more detail in Chap. 2), and secondly, people can become cruel if the social norms, the social and political system, and the situation allows this. I hope to have now generated a greater interest in the topic of prejudice, and readers can't wait to see what might happen in the brain when someone is feeling negative about an out-group, and also if we can and should ever cure prejudice.

This book is interdisciplinary, and it will cover several different scientific areas. Specifically research in neuroscience, psychopharmacology, psychiatry and philosophy are included. In Chap. 2 'The foundations of prejudice and discrimination' I will start with discussing traditional textbook social psychology accounts of prejudice and the latest social psychological research about prejudice and discrimination. This is not boring, nor is it easy; often when I ask my students what prejudice is, they do have *some* idea. In fact everyone knows the word. But in order to understand it fully, it requires more than just knowing the word and having some vague idea what it might mean. Is it an attitude or a feeling or is it our knowledge? When is someone prejudiced? In todays' modern society people are very careful not be seen to be prejudiced. Is it prejudice if I prefer to be around men rather than women? Secondly in Chap. 2, I will describe another concept; that of implicit negative bias. Research has shown that even people with a sincere belief in equality have some unconscious bias, that makes them prefer some groups (their own), over others. I will describe a very popular test which the reader can take on the internet to measure their own unconscious bias against other races, ages, religions etc. Chapter 3 'The Neuroscience of intergroup relations' will then describe the neurological basis of prejudice. When I say neurological, I mean that, of course, prejudice has a basis in the brain, as we are our brain. I will come back to this later in the book. Sadly, there is no reason to believe in a soul outside the brain for most neuroscientists. Every feeling, every character trait, every experience, and every thought is in the brain. So also our prejudice is in the brain. This does not mean that we are necessarily born prejudiced. Regardless of the origins though, all is of course routed in the brain. The brain is still poorly understood as it has an uncountable number of network connections. Later, I will describe what happened in experiments where people were shown pictures of members of other groups while their brains were scanned. Also, I will describe what methods are available to investigate brain function, and what this research can tell us about the nature of prejudice.

Core to this book, is Chap. 4. "Cure prejudice"? Does that mean prejudice is a disease? Cure with what; a medicine? Surely that is not possible? And it will never be possible, right? In 2012, we published a study in the scientific journal Psychopharmacology. This study was reported world-wide in the media; mostly with this heading: "Cure for racism found.", or: "Take a pill, and change who you are." I will describe this research and the implications for neuroscience research on prejudice. Chapter 5 then describes how some forms of extreme persuasion (brain-washing) might also change the brain. What happens to our brain in those cases? Can the environment and persuasive messages lead to long lasting changes in the brain? I will discuss how methods of 'changing someone's mind', are relevant to a consideration of cults, marketing, torture, and social influence and how non-medical 'interventions' might also change brain networks. In Chap. 6 'What should be done?' I will describe philosophical and ethical debates. Here I will address the question; if one could cure prejudice, should one? After reading the scenario at the beginning of this chapter; about the seemingly nice guy, who then turned out to be a Nazi concentration camp worker, and considering that what he did was partly caused by his strong prejudice against Jews, people might want to immediately say "yes, if there was any way of reducing his prejudice then do!" Or would some people think differently? What about 'curing' individuals who commit hate crimes, fuelled by their prejudice? Should we force them to change if we could? In UK criminal law there is a dispute: for instance, if someone is treated in a high security psychiatric hospital for diagnosed paedophilia, the person still has the right to choose whether to take or refuse a drug that reduces their sexual drive. In this case often there has been a confirmed criminal offence, which was also caused by a recognized mental disorder; however the person is not tied down and forced to take the medication; they can choose. What then about Nazis? And if there was a drug that would prevent them being prejudiced and aggressive, should we just give it to them? This might lead to the quite disturbing scenario of a future in which everyone was taking drugs, or changing their brain for the better or worse. Indeed, at first glance this all might sound like some mad scene out of a science fiction book. In fact, there are some philosophers and ethicists who would argue that one should accept small sacrifices if it is for the larger benefit of society. This is what is often referred to as the utilitarian ethic. Numerous papers have discussed cases of performance enhancement; taking drugs to be faster, quicker, smarter. Why not do it? Or is it more complicated than that? If social influences can change the brain why not medically intervene? I will address all these questions, and also summarise a recent publication where we discussed arguments for and against "society's moral enhancement". Besides this question, there is however also the question of whether people really want a society where there is no prejudice, where everyone is treated as an equal? Surely we are allowed to draw some lines; would it not be morally permissible to help *your* drowning child over a stranger, because your child is related to you? But where do we draw the line? Is there a cure for prejudice? And if there is or will be in the future, should we use it?

Open Questions Chapter 1

- Do you think people today could behave in same manner than they did during 2nd world war in Germany?
- Do you think the fictional character from the beginning of the story is mentally ill?
- Do you think you could kill anyone if there was no law against it?
- How could such events have been prevented?

References

Atran (2003). Genesis of suicide terrorism. *Science, 299*, 1534–1539.
Swanson, P. (2015). *The kind worth killing*. London: Faber & Faber.

Chapter 2
The Foundations of Prejudice and Discrimination

Most people recognise prejudice when they see it. For example in the fictional story – 'Roll of Thunder; Hear my cry' – in which young black children in 1920s America describe the terrible conditions and the unfair treatment that they faced in their everyday life, a record of a school book that children borrowed is shown. Here people might see the obvious discrimination; as the condition of the school book has deteriorated before the black student is given it. Thus race can be regarded as the key factor in the decision of who gets which book. Multiple works of fiction deal with the problem of prejudice and discrimination and describe the experiences of those that face such unfair treatment. For example, another book that describes the experience of black American children is the artistically written famous novel "Song of Solomon" by the winner of the Nobel Prize for literature Toni Morrison (one of Barack Obama's top 10 favourite books) (Morrison 2006). Besides the American history of apartheid we can see other more extreme cases of prejudice at times of war, when prejudice fuels the perpetrators desire to kill out-group members – the enemy -. This too is often depicted in works of fiction, and most dramatically in first person accounts of war victims. Sometimes such accounts, even when presented fictionally, can indeed be very distressing to read. 'The Storyteller' by Jodi Picoult is a novel which describes the 2nd World War experiences of a young woman (Picoult 2013), and Ken Follett's 'Winter of the World' follows different families during the 2nd World War (Follet 2012). Prejudice and discrimination are not only part of history; we see in the daily news, how individuals suffer from prejudice and discrimination. The Independent newspaper reported on 15.03.2015 about a Syrian refugee mother – Hanigal – living in poor conditions with her 15 year old disabled child. In Syria, 200,000 people have already died, and Hanigal is among 1.6 million who have escaped the terrors of conflict. "At first, I was living on the streets, with nowhere for my child and me to go." Hanigal said. Racism is not only prevalent in the extreme conditions of war, for example in football there are still accounts of racism, as described in the Guardian newspaper, and also in Emy Onuora's book "Pitch Black: The Story of Black British footballers" (Biteback Publishing, 2015). On 20.08.2015 the German magazine "Der Stern" published an interview with a

© Springer International Publishing Switzerland 2016
S. Terbeck, *The Social Neuroscience of Intergroup Relations: Prejudice, can we cure it?*, DOI 10.1007/978-3-319-46338-4_2

member of the group Ku-Klux-Klan, which illustrated how racial prejudice and hatred still prevails in contemporary America. Richard Preston lives in a little wood cottage in South Virginia. Outsides his house he has several banners, one stating for example: "Rebel brigade – Knight of the invisible kingdom.", as well as one reading: "For God, family, races, and nation." In the interview with Stern he said that:" This country (America) is at its end. Our wives are raped, white men are attacked, and Christians are killed. But we will not let this happen." Indeed, this might also illustrate the fear – fear of loss of territory – that this man must feel, and that others might not feel. Going out with burning flags he shouted "White power". In his book "Them", Jon Ronson (Ronson 2001) describes multiple interviews he held with extremists. He spoke to believers of conspiracy theories, as well as right-wing extremist. Again, it might become clear that a simplistic idea or a black-white world view might underlie some ideologies. For instance, this was a conversation between a member of the Aryan Nations and Jon Ronson: "The Anti-Christ Jew", he said: "The same one that murdered Abel." "All Jews, or just some Jews?" I asked him. "All Jews!" he said. "It's a blood order, DNA has proven it." The Stern magazine noted that the threat posed by extremist groups in the USA was previously largely underestimated. Recently, a US policeman was caught on film firing at a black teen-ager 16 times, involving many times when the teen was already on the ground (Metro, 26.11.2015).

That prejudice still prevails can also be seen in behavioural experimental tasks, for example one which is entitled "shooter task". In this computer task animated black and white avatar males appear in the background either holding a gun or an innocent object. Researchers found that Caucasian subjects were more likely to 'shoot' unarmed targets in the game if they were black. In 1940, African-American psychologists Prof Kenneth Clark & Prof Mamie Clark conducted the well-known "doll-experiment". In this experiment children were presented with two identical dolls, except of their skin and hair colour. The children were asked: "Which doll would you like to play with?", "Which doll is the good one?" "Which doll is the bad one?" The majority of Caucasian children preferred the white doll, on all accounts, wanted to play with it and stated it was good whilst the black doll was bad. This effect was strongest when the children were in segregated compared to mixed race schools. In 2005 filmmaker Kiri Davis repeated the doll study, as part of his film "A girl like me". Sadly, even in 2005 – nearly 70 years later – Kiri reported that he found the same results as in the original study, that children strongly preferred dolls of their own race (i.e., white children preferred the white doll, black children preferred the black doll). Also they found that children wanted to play with the own-race doll more, and more importantly, that they also thought the other race doll was "bad".

However, the above are only a few examples, where most people would agree; this is prejudice, discrimination, racism, unfair and morally wrong. At the 2015 European Congress of Psychology in Milano there were many researchers from nearly every country of the world. Many were from Italy and Europe, but also America, China, Japan, Brazil, South America, Australia and New Zealand. In one 2 h session on prejudice there were six short talks given by people from different countries. There was a talk from a researcher living in South Africa in which she discussed prejudice between urban blacks and rural blacks. Then I heard a talk from a researcher living in Turkey, he reported that Syrian refugees face prejudice from Turkish people. A researcher living in Greece found that Albanian immigrants face prejudice from Greek people. There is prejudice and discrimination everywhere. Indeed, there might even be "prejudice" against an unknown group or non-specified group. For instance some people worry seriously about lizards ruling the world others might be concerned about the actions of 'an establishment' 'The New World Order' or, like KKK member Richard Preston stated, fight for an "invisible kingdom".

In 'The Psychologist' magazine (published by the British Psychological Society), in July 2015, Paul Guhman writes about the caste-based prejudices, which even affect British Indians, living in the UK. In particular, he discussed how the traditional Indian caste systems even prevails amongst Indian individuals living in the UK. For instance "the untouchables" (called "Dalit" or the oppressed in the Sanskrit language) are still at the bottom, in terms of housing, education and social care. Dalit encompasses all people outside of the caste system and Chandalas has a very specific meaning; the latter deals with the disposal of dead bodies only, and whose mere touch could contaminate the upper classes. Guhman (2015) described how the Indian caste system maintains itself though endogamy (intra-caste marriages only), separate places of worship, early socialisation with kinship, as well as caste-based community centres. He reported that people considered castes as hereditary, hierarchical, and justified in Hindu scriptures and traditions. Guhman found that children as young as 7 years of age knew not only about their ethnicity, but also their rank-order (or caste) they belong to. Indeed, the phenomenon to favour one's own group has been observed in every country in the world. For example Mikey Walsh's autobiographical "Gypsy Boy" (Walsh 2010) describes prejudice from non-Roma schoolchildren against "gypsy" children and also Roma children being prejudiced against non-Roma children; "We were always conscious of them watching us through the gaps in the trees, but were warned not to ever speak to them. 'Gorgibreds', our mother would say. 'Don't you ever speak to them, even if they talk to you. They'll have you taken away.' The prejudice went both ways. 'Come away from there' we'd hear their mothers say as she shoved them back into the house. 'They're Gypsy, and they'll put a curse on you.' One day Frankie I and heard the girls whispering; 'Gypsies, look it's the Gypsies.'"

In the past, racism was often casually and unquestioningly accepted. Nowadays, at least in some places and with regard to some prejudices, people are increasingly worried about appearing prejudiced. For example, in research today we sometimes find that when asking how much money one would consider giving a black or white person (with the participant being Caucasian) participants would not vote for an equal share, but consider giving the black person more. This might also indicate that some participants were worried about appearing prejudiced, thus being extra generous (i.e., positive prejudice), even though it might also be that they genuinely felt it was fair to give the minority group more. In the current western climate prejudice and racism seems to be a very sensitive topic, and many people have a great fear not to say anything "wrong" that could class them as being racist. On 21st of August

2015 the German local newspaper "Emsdettener Volkszeitung" reported on an incident regarding a German TV program called "Aktenzeichen XY". Similar to the US TV program 'Most Wanted', in this show recent unsolved real crimes are reported. Often a picture of the suspect is shown, asking the audience to contact the police if they know this person or have any other information related to the crime. For the next German program it was scheduled to report about a rape case, but the program directors had decided to not report about this case in the show, as the suspect in the crime was black, and this might increase prejudice against black people. However, on 22.08.2015 the newspaper reported that now the program organisers had changed their mind about this decision and finally decided that they WOULD actually report about this crime, as they said that the mere mentioning of the ethnicity might not be seen as racist. This might illustrate the level of anxiety and insecurity that individuals feel when discussing issues about ethnicity. This is also highlighted in a recent UK Channel 4 TV documentary, in which Trevor Phillips, the former head of the commission for racial equality, discussed "10 true things about race you can't say". For example one of these things was that by mere statistics, Romanians in the UK are more likely to be pickpockets. On 15.01.2015 the Daily Mail newspaper published an article entitled "Branded racist at five". According to this article schools teachers are reporting primary school children for using the "wrong" terminology when talking about other students. For instance one boy said he wanted to play with the "Chinese boy", as he did not know the boys' name. According to the newspaper, the teacher reported this incident, explaining that one should be addressed with their name rather than their nationality. Thus, it seems an important task to define prejudice, stereotype, discrimination, and racism, in order to see what one is (and is not) talking about.

What is prejudice? What is racism, what are stereotypes, and what is discrimination? One problem with social psychology might be that we sometimes use terms, theories, and ideas that at first sight seem easy to understand. Asking 'what is prejudice?' is not like asking "What is the phi coefficient?" To the latter question there is one answer, and someone without a background in the field of mathematics, who does not know the exact definition of phi coefficient, simply does not know the answer. With prejudice this is different; everyone could *somehow* define it, but without using the exact social psychological definition of it. That can be a problem, because when talking about prejudice you and I might have a different definition of what we are talking about. Sometimes, students tend to answer exam questions in social psychology in the same manner. For instance the question "Why do we have norms?" I did not mean students to just have an educated guess about this question, but to obviously mention the social psychological theories from the lectures. In fact I felt quite the same when I was a student of social psychology; I thought this subject was blindingly obvious. For instance my lecturer at the time told me about a social psychological study; he said that a large group of researchers in social psychology conducted an experiment; the researchers found that participants are more likely to give money to a friend rather than to a stranger. I thought: "Hang on, that's it? And they did an experiment on that? I could have told you that before". At first glance then this might lead to the idea that social psychology was just common knowledge, described in complicated terms. However, this is not the case; it is important to understand – really understand – what researchers are referring to using seemingly complicated terms. What do they mean when they say "attribution error"? What do they mean when they say "fairness norm"? What do they mean when they say "prejudice"? I believe that this is the key to actually becoming a social psychologist, to see that the definition and the theory behind the terms are essential. If one understands what previous researchers have meant when they defined certain social psychological terms then one can understand what seems to be merely simple ideas in a deeper way. This understanding then also leads to finding out what might be wrong or problematic with the current definition and this allows ideas to develop and progress to be made.

I will now discuss the social psychological definition of prejudice, racism, stereotypes, and discrimination. In fact, before I do this, there is a second reason as to why it is important to actually truly understand what the terms mean, and that is a problem related to research methods. Put simply; 'If you don't know what it is you can't measure it'. In psychological science the key is to measure variables. Take the example that psychologists found that watching violent films increased aggression. How could you do an experiment testing that? First you'd have to define aggression before you can measure it. Is it physical aggression? Is it verbal aggression also? Hitting? What about aggressive sports? What about a person that wants to be harmed (a la Fifty Shades of Grey)? So try a definition of aggression; Aggression is harming someone (physically or verbally) who does not want to be harmed. But now what about aggressive thoughts? What about harming properties? It becomes clear that the definition of the concepts already determines how I am measuring it. Say there was now a definition, how can one measure aggression? How can one measure if someone was aggressive? With a questionnaire? But participants might not want to admit being aggressive. Observing them? But they might not act aggressively even if they feel it. And coming to that; what are violent films? How violent? What violence will be depicted? For how long are participants watching the film? Do they have to watch it over again? Were the participants aggressive to start with? Which gender? How can you ensure they are really watching it and not finding it boring? Is it important if they like the film or not? Indeed, I now described a seemingly simple experimental finding "Watching violent films increases aggression", and I can name numerous problems with that. And then the seemingly obvious facts seem all the more complicated. And that ability, to question everything, I think is what makes researchers able to develop new ideas. Back to prejudice then.

In social psychology prejudice is defined as an attitude towards a group of people. Indeed, as one has attitudes towards objects one has attitudes towards people. For instance saying: "I have a positive attitude towards sports." basically means "I like sports". Or I have a positive attitude towards food, means "I like food." I have a negative attitude towards illegal drugs, means "I don't like them". Thus, if we are not talking about objects but people, having a certain attitude towards a group of people is defined as prejudice. And indeed, there is also positive prejudice, e.g. a positive attitude towards a group of people. So, even though mostly negative prejudice is studied, by definition prejudice is not negative, but can be both; positive or negative. Although, negative here does not mean, it is negative, therefore it is morally wrong, it might be that as well, but more on that in the final chapter. Here, it simply means that the person has a negative attitude towards a certain group of people. i.e., they don't like them. For example, most people might have a negative attitude towards paedophiles. Often I also find in my research that most people report that they have a negative attitude towards drug addicted individuals. Regarding race, age, gender or disability, being prejudiced generally means having a (negative) attitude towards that group.

Now I need to describe what attitudes are, in order to understand the latter sentence more. Often researchers suggest that attitudes have an emotional and a cognitive component. I will illustrate that with objects first; Say I have a positive attitude towards sports, then every time (or most of the time) someone mentions sports I will both; feel good to hear about it (the emotional component), and also think positively about sports (e.g., "Sports are really good for me. Sports make me fit.") (the cognitive component). Now am I actually doing sports? Well that's a whole different question. That is the question about the relationship between attitudes and behaviour, which is actually quite weak. So taking the extreme case, I could well have the most positive attitude towards sports, really enjoy hearing the word, thinking it's good for me, thinking it makes me fit, but actually never do it (more on that later). The basic idea is that an attitude towards an object or a person involves thoughts as well as emotions and feelings when one encounters that object or person. Indeed, the same logic applies to prejudice. As one sees a person from that group certain thoughts and feelings arise. For example one could feel fearful or aggressive when they see a person from that group. Or one could also think; "That group is not very good", "They are bad for me" etc.

Reading this, one might become puzzled wondering that it must be almost impossible NOT to do this, not to feel or think anything when they encounter another group. You have to have a positive or negative attitude towards a group of people? Or could you have no attitude? Or could you have a neutral attitude? I think a key problem here is the question of why we form these groups. Why do we group people together based on race? With objects or activities it is obvious; my attitude towards sports, my attitude towards food, but why not my attitude towards people with green eyes? Why race? Why gender? That is indeed a key problem. And maybe a bigger problem than prejudice is the problem that humans have a tendency to group – or categorise – people. Is there nothing we can do about that? Thus, before talking about prejudice, stereotypes, and discrimination further, we need to think about something much more basic; the formation of groups, more specifically, the formation of in-groups and out-groups. Sometimes people assume that 'out-group' has a negative connotation, however by definition in-group means the group the individual belongs to and out-group the group the individual does not belong to. This can be in terms of race, gender, age, religion, political option etc. So for instance, I am female (females are my in-group) this creates my out-group; males. But why do people form groups? And why do they form groups based on race, age, gender, and not on eye colour, nail biting (or not), long-short hair etc.? So, one key problem then seems to be that we form groups, and that we treat the group as a collective and homogenous. Indeed, if we did not have the tendency to form groups there would be no prejudice.

Humans may have a natural (probably universal) tendency to form groups. Have you heard about the Gombe war? It lasted 4 years, from 1974 to 1978 in the outskirts of Tanzania. January 1974 marked the outbreak when members of the Kasakela (the northern subgroup) attacked and killed a member of the Kahama (the southern subgroup). In the following years, further females – now Kahama – were killed, raped, and kidnapped. The war ended with the Kasakela taking over successfully the territory of the Kahama. Does this not sound very human? In fact this war was a war between chimpanzees. Primatologists have long wondered and researched such observations and the questions "Why would chimps kill each other?" In the journal *Nature* Michael Wilson from the University of Minnesota published a review and long-term analysis paper, in which he included the combined observation of many different chimpanzee communities across Africa, including over 426 primatologist observational studies. Using a computer model to investigate the cause of violence led the researchers to the conclusion that the most likely explanation for the chimpanzee violence was evolutionary adaptive strategies, in particular the formation of in-and-outgroups. Indeed, most animals that chimpanzees attacked were animals outside their in-group. In his book on geography and anthropology, Diamond, J. (2012) illustrated the social lives of tribal societies in North and South America, in Africa, and in Australia. He stated that traditional societies deliver the opportunity to study social behaviours over thousands of millennia; natural experiments of the human race. For example he described visiting a mountain village in New Guinea where the people living close to the river described their group as friends, and the people living at the mountain as the 'bad mountain people', the enemies, who were further described as evil and subhuman. Diamond described how he heard from a Wilihiman Dani man in New Guinea living in a tribal society:" Those people are our enemies, why shouldn't we kill them? – they are not human". Diamond expressed his surprise as both groups to him looked the same, spoke different, but related languages, but understood each other's languages. Diamond suggested that in tribal societies, if one was to encounter a stranger then they would have to presume that this person was dangerous, because the stranger would indeed be likely to kill people in his clan, and try to invade their territory. Thus friendship would only emerge within one's own group, which was mostly just ones extended family. In somewhat larger societies however, Diamond argued, business exchange and mutual supply were the first steps of positive encounters with members outside one's own group. Besides discussing tribal human societies Diamon also investigated animal behaviour and discussed animal species that engage in in-group/outgroup and war-like activities, animals such as lions, wolves, and chimpanzees. He suggested that two features distinguish animals that engage in war like behaviour from those animals that don't, which is competition and variable group sizes (e.g., it was safe for a large group of animals to attack a smaller group and steal their recourses). He suggested that animals as well as humans might have been predisposed to engage in pro-social as well as anti-social behaviour. The circumstances (such as resource limitation, group size) then determine if they engage in war or peace.

But why do we have the tendency to create in-groups and out-groups? In 1977 Sherif conducted a social psychological experiment which also became known as the Robbers Cave boys camp study. Boys from Oklahoma were invited to a summer camp. Arriving at the camp the boys got to know each other, played with each other, and enjoyed their time. Then the experiment started; the boys were divided into two groups. Now they played competitive games, one group against the other. Sherif wrote that "If an outside observer had entered the situation after the conflict begun... he could only have concluded that on the basis of their behaviours that these boys were either disturbed, vicious, or wicked youngsters" (Sherif and Sherif 1969, p.254). Indeed, creating random groups created more and more suspicion and hostility. The boys would vandalise the property of the other boys group, steal their possessions, and play war. What happened there? How could these nice boys have turned into being so anti-social? In a different study school teacher Elliot created 'A class divided'. She divided her class by eye colour; children with blue eyes, and children with brown eyes. She also told the children that those with blue eyes were smarter, nicer, neater, and generally better than those with brown eyes. There were two key observations to this study. Fist, the teacher was able to create hostility and biases amongst the children. And furthermore, subsequently the blue eye children did really perform better at the tests. The above studies suggest that in-groups and out-groups can be created with random attributes, (you are group A, and you are group B), but then why do people categorize according to race and gender, and not eye-colour normally?

Leslie (2015) suggested that humans are inclined to generalize from experiences, and that human's categorize objects as well as people into categories that share hidden properties. The authors suggested that especially if the attribute is negative then humans perform what they call a "rapid generalisation". For example they state that one does not wait to see if all tigers bit, a single instance might be enough. Most importantly however the authors discussed the question of why people then categorize according to race, and not normally to eye-colour. They argue that the categories to which people generalize attributes is learned in early socialisation, for example from parents, peers, culture and media. Thus, the traditional view that social categorisation was mostly based on visual cues, such as race, age, gender, is challenged since more recent research has demonstrated that prejudice can also arise from different characteristics, such as political orientation.

Berreby (2005) argued in his book on the social psychology of intergroup relations, that the idea that people usually perceive others just the way they are, using 'true' categories, such as age race and gender, was wrong. He suggested that there were perception biases in social cognition. Thus "...the issue is not what human kinds are in the world, but what they are in the mind." He argued that humans are much alike, but that they are also different, and that taking any random group of people could create a group that was distinctive somehow; For example the women on the boat, or the five men in the café etc. Berreby also argued that categories would change over time. For example in 1400s France there were people which were recognized as a distinct human kind, called cagots. At that time people belonging to this group would use different entrances to, live separate lives, marry apart, etc. Indeed, cagots were ostracised by society. However, centuries later, this category was not used at all anymore, it disappeared, and cagots were re-categorised. Furthermore, researchers also found that during and after the 2nd World war, American and Japanese student's descriptions of the other nation's traits had changed. For example whilst before the war Japanese were described as progressive and artistic, they were then described as sly and deceitful. Stereotypes might be created by what Berreby called the 'looping effect'. This means that an idea is formed in someone's mind that a category of people are onto something, and others get persuaded, and the idea spreads. This then leads others to use such attributes too to guide their behaviour and decisions. Berreby gives the example of humans being able to see human in motion even when there are only a few dots moving, thus suggesting that people perceive significant differences, even though these may be often illusionary.

Researchers have conducted an experiment in which they presented participants with moving shapes. Participants were quick to gives those shapes human characteristics, such as "the large red box is chasing the blue square. The blue square is anxious and running away. More recently, Paul Bloom and Csaba Veres from the University of Arizona created a new version of this experiment, in which they instead had groups of shapes moving. Participants interpreted shapes moving together as a group, and also associated the characteristics that they attributed not with the individuals but with the group as a whole. This suggests that social categorisation happens in the brain and not in the eye. The key point Berreby (2005) was arguing was that: "The important thing to understand about human kinds like race, ethnic group, nationality, and sexual orientation is not that they're baseless. It's that they make no more sense than alternative categories, which we do not use. That's how we know that the source of our beliefs is not physical evidence of people." Indeed, both neuroscience as well as behavioural research suggests that at least it might be possible to redirect the focus of the perception of an individual. For instance, as our own research shows, changes in basic emotional arousal might indeed change whether one sees race or not (to the same extent). Furthermore, the initial categorisation might distort future perception. For instance I remember an incident when I saw a nice handbag in a shop but did not buy it. Subsequently, I suddenly looked at – or noticed – everyone's handbags. Similarly, you might have the experience that if you are about to go to the dentist, suddenly you notice everybody's teeth, or if you had a haircut, suddenly you notice everybody's hair.

Kurzban et al. (2001) supported the idea that race might not be always a crucial category for in-group out-group distinction. He conducted an experiment in which different teams were brought together. They had two attributes, some were of different race, but rather than race, t-shirt colour however predicted team membership. The authors then found that t-shirt colour was more strongly encoded than race, supporting the idea that encoding *coalition* was the most important. A further study from the University of California, Santa Barbara investigated whether political orientation could reduce classification of individuals according to race. Participants listened to Republican and Democratic arguments. Those who defended Republican views contained black and white people and the same for Democratic views. The researchers found that the superordinate category (political orientation) reduced the classification according to race. In a news article Prof Todd argued that human brains were not designed to attend to race, but were instead designed to attend to coalition. He said that:" Race gets picked up only as long as it predicts who is allied with whom". The 2nd world war novel "In Love and War" Preston 2014 also describes a scene – played in Italy – which illustrates how coalition might outweigh nationality or other factors. "'We need to make things straight', he says. 'I like no English. We not need English here. For too long English treat Florence like a home. But' he gives a reluctant grin 'Podesta says you good Fascist. For me nationality is not so important, Fascism is important.'" Indeed, also Greene (2003) suggested that morality evolved to promote cooperation, but cooperation within groups (any group, even as small as two individuals) creates competition between groups. So, is there any additional psychological function that maintains groups once we have formed them?

Tajfel and Turner (1979) developed the social identity theory. This important theory suggests that an individual not only defines themselves as an individual, but also as a member of certain groups, e.g. as female or male, nationality, religion etc. Social identity theory (Turner et al. 1987) further describes the fundamental influence of how people categorize themselves. For example under some conditions their self-concept might shift from seeing themselves as an individual "I", to a social or collective level "we/us". Besides this, people also define their own status by group comparisons. How do you know that you are pretty? How do you know that you are rich? Because of social comparison. An individual compares themselves to others also to obtain information about their own value. If I find that I earn less than anyone else, than I know I am poor. Thus comparing oneself to groups and making a positive distinction (i.e., being better than them) also increases one's own self-esteem and value. So reducing the impact and value of other groups also increases one's own value.

I will now turn to the third concept that might be associated with 'prejudice' which is stereotypes. So far I have discussed prejudice and group formation (ingroup and out-group). Stereotypes is now the third concept that needs precise definition, as indeed this concept might also be misunderstood. According to the social psychological textbook definition a stereotype is the knowledge (correct or incorrect) an individual has about a group. Germans drink beer. Italians eat pasta, nurses are caring, etc. These last three stereotypes I mentioned are not as controversial as others. As discussed earlier, in a recent Channel 4 documentary, it was proposed that there might be certain things about race that one could not say but were true. A problem here might be the presumption that knowing about the stereotype equals racism, which in fact is disconfirmed by much research in social psychology. One study it was found that individuals' showed levels of prejudice, independent of how much they knew about the stereotype. For example one might know all apparent stereotypes about a certain religion or race, but not be prejudice at all.

In his famous book 'The nature of prejudice' Gordon Allport (1956) argued that many stereotypes might have a kernel of truth. Recently, someone presented a large study to me, which apparently showed that women – on average – had a lower IQ on some intelligence tests than men. Of course this might make me a little upset (as I am a woman!). But looking closer at the study I determined that women had a significantly lower variety of scores, clustering around the average, whilst men were either very intelligent or very dumb. Looking even further I determined that the average difference that they apparently found was 4 IQ points (which is tiny). It might be that results would have been different if they controlled for motivation, or extraversion, or if they used a different IQ test, and most importantly, controlled for age. As with many experiments, it is hard to make any such presumptions. But it should also become clear investigating if some stereotypes are partly true, or not, or under certain conditions, etc. etc. does not help the debate on prejudice as the knowledge of the stereotype seems to be independent of prejudice and indeed, discrimination. What can create a problem however is, if a person is just described in terms of their group membership. And in particular, if only one attribute (e.g., gender) and nothing else is considered. Thus individuality is overlooked, which indeed goes back to the problem of forming in-groups and out-groups, which then leads to the development of negative attitudes (i.e., prejudice) against individuals. Thus when I hear an argument, such as;" On average woman are less intelligent. I don't believe *all* woman are less intelligent, but if I had to estimate IQ and gender as the only information then I could use that to judge a person, as it is a reliable predictor.", I disagree. I disagree because in real life one never has nothing else other than the information about gender available (name any real life scenario in which one would just know a person's gender, but not also their age, attractiveness, etc. etc.), and thus relying on one attribute is not reliable as it overlooks the others. For example to judge a person's intelligence just by their gender (even if one were to agree that the stereotype of gender was true), is not good because it's insufficient; age, education, motivation and many other characteristics might be important predictors.

Finally, it is important to note though that prejudice, indeed even very strong prejudice, does not necessarily lead to discrimination. For example there might be a person who is very biased and racist, but does not want to show this, so in fact never overtly discriminates against someone (although I have to admit that seems unlikely). Research in social psychology has shown that indeed, there is a relationship (although not that high) between attitudes and behaviour. And indeed, this is I believe, the most problematic – and also morally wrong – aspect, to not treat every individual as an equal, or more precisely as a moral equal. Philosopher Gracia defines racism as "… what makes someone a racist is her disregard for, or even hostility to, those assigned to the target race … she is hostile to or cares nothing (too little) about some people because of their racial classification … hate and callous indifference (like love) are principle matters of will and desire: what does one want, what would one choose, for those assigned to this or that race?" The Gracia (2004) author furthermore suggests that cognitive (related to justice) as well as emotional (related to care and benevolence) aspects are involved in the racism definition.

At the beginning of this chapter I started by explaining that we recognise prejudice when we see it, I took the example of a schoolbook, which was given to white children when it was new and to black children when it was used and no longer in a good condition. That is overt discrimination. As we have seen at the beginning of the chapter, it might not only be prejudice that leads to such behaviour, as someone who was very prejudiced could also not act upon their believes and feelings. And on the contrary, someone who was not prejudiced might act in a discriminatory manner, for example if the social norms of the time or country pressure individuals to certain actions, or if the political structure in the country is insecure or if rights of individuals are not protected etc. Thus, as I also discussed in Chap. 1, the biggest problem that society might face, the problem that not all individuals are treated as a moral equal, might only partly stem from prejudice, but also from other social and political factors. Thus, even if there was a drug to reduce prejudice, there might still be discrimination, hostility and war.

Within western democracies partly social and moral development and education has led to societies in which most people do not overtly discriminate against someone because of their race, age, gender, or ethnicity. In fact in most western countries there are laws against such behaviours, so that legal charges would follow if someone was overtly hostile and discriminatory. Even though, we have cases of overt discrimination still reported in the news, overt hostility is not the norm. But what is often observed though is what is called; subtle discrimination. One is just less friendly, less close to someone from an out-group. They might also be less willing to offer them a job, or more likely to notice their mistakes, or they are less willing to choose someone from an out-group as a friend or partner. Even though one could argue that this form of discrimination is not as bad – after all no-one gets killed – it might be very dangerous to assume this, and it might also be wrong to ignore the harm that can be caused to an individual even if it is sometimes not physical harm. As I already discussed in the introduction, it might be wrong to assume that just because we do not have an acute war situation that discrimination does not exist or that it is marginal. Previous research for example found that racial biases can occur in hiring decisions. In a study in America, fabricated CVs were sent out to job adverts, with some of the applications using a stereotypically black name for the applicant and some using a stereotypically white sounding name (such as Greg and Emily versus Lakisha and Jamal). White sounding applicants received 50 % more call backs for interview than applications from black sounding applicants. Furthermore employers who listed themselves as an 'Equal Opportunity Employer' discriminated as much as employers that did not describe themselves in this way. Is that fair? Do we not want to get the best qualified applicant regardless of race or gender?

In social psychology there have been many behavioural experiments investigating methods that might help to reduce prejudice. The most cited and the most well-known is the contact-hypothesis. This hypothesis was developed by Gordon Allport in the twentieth century. It is postulated that contact with people from out-group can reduce prejudice. Later, it was added that it can't just be any contact, but that certain conditions needed to be met such that the individuals needed to be of equal status and were working towards a common goal. However, in a recent meta-analysis (a comparison of a large number of studies conducted on this topic) it was concluded that contact did not even have to meet those criteria, but that any contact helped to reduce prejudice. The theory has triggered many studies around the world, with policy implications, such as increasing the number of mixed background schools, supporting mixed neighbourhoods etc. The social psychological studies investigating this topic are mostly questionnaire based (i.e., surveys), but there are also experimental studies. In the questionnaire based studies participants are asked about their attitudes towards out-group members and they are also asked about quality and quantity of contact they had with members of that group. What researchers find consistently is that those with high contact also have more positive attitudes towards outgroup members. Now one could argue that this correlational relationship might also mean that those with positive attitudes to start with are searching for more contact, and that it was not the contact that improved the attitudes. However, experimental studies – in which you bring groups together rather than just post-hoc assessing their contact, support the theory that indeed the contact causes an improvement in attitudes. Furthermore with a large sample it is possible to use certain mathematical procedure (i.e., Structural Equation Modelling) that enables making causal rather than correlational claims, about the relationship. That contact reduces prejudice has not only been confirmed in western democracies or in areas where overt prejudice was quite low, but also in high conflict areas, such as Northern Ireland, and places within Africa. In experimental studies, participants are usually brought to the lab and then they either meet a member of the other group, or they can also merely imagine meeting someone from the other group. Participants are then asked about their attitudes towards that group before and after the contact. In conclusion a large number of studies support the hypothesis that intergroup contact can reduce prejudice. But this does not tell us yet why? Some studies have suggested that contact reduces intergroup anxiety or that contact increases empathy towards to other groups or that contact reduces perceived boundaries between the group (i.e., they are less likely to be perceived as members of an in-group or out-group, but rather as more similar), or a combination of all of those effects. Interventions to reduce prejudice could tackle a variety of factors; for example contact could increase positive feelings (or reduce negative feelings) towards the out-group, or it could increase empathy towards the group, or it could lead to re-categorisation or de-categorisation.

As I have discussed before, if people did not have the tendency to form groups, there would be no prejudice, so if an intervention reduces people's readiness to sort humans into groups or categories then this would reduce prejudice too. Another set of factors that have often been linked to reducing prejudice are interventions to increase empathy. Empathy, different from sympathy, involves feeling the other persons emotions to some extent too. For instance, when a child is crying one could feel sorry for the child, which is sympathy. Empathy involves the individual partly also feeling as the child feels (sad when they are crying, angry when they are angry, pain when they are in pain). In the social lectures I usually illustrate the concept of empathy by telling my students: "In one minute you will all feel empathy. Guaranteed" Then I darken down the room, showing nothing, and then I show the image of a hand with one finger that has just been cut off. "Ahh" all students would say. "See, I told you" I would say; "You feel their pain. That's empathy" Indeed, many fMRI studies (fMRI study methodology, will be described in Chap. 3) have shown that when seeing someone else in pain, some of the same brain areas are activated as when participants have the pain applied to themselves.

fMRI studies have also determined however, that the race of the person is a significant factor in this effect. If the person being viewed in pain is from ones' own group the empathy network is more activated when compared to viewing an out-group member in pain. This suggests that people feel less empathy (for pain) towards out-group compared to in-group members. Interventions to reduce prejudice might thus try to increase empathy towards others. How could this be done? In experimental settings there are a number of ways in which researchers have successfully increased empathy. One way to induce empathy is to simply give participants the instruction to imagine how someone in a particular situation might feel. This is something that is also often used when people try to increase empathy. For instance in a school setting a naughty kid that has just hit another child might be told off by the teacher and then also be told to: "Imagine how you made the other child feel". Or within romantic couples, when one partner is late for dinner the other partner might say "Do you know how I felt waiting for you?" In both cases one is addressing the feelings the other person might have had. However, there is another – more cognitive concept – related to understanding the other person, which is perspective taking. Perspective taking involves rationally understanding the other person, which might or might not lead to empathy (feeling similar emotion to the other person). For example if someone fell over and cried, one observing this could cognitively understand that this person was in pain and would therefore cry. However, they might feel no empathy for this behaviour which means that they would partly feel their pain as well. Sometimes movies that don't successfully induce empathy illustrate this effect quite well.

We observe an actor being beaten by a gangster, the person cries, but we don't feel upset (as the movie is not well made and we don't feel for the actor), even though we logically understand what is happening. One key scientist researching empathy is Prof. Dan Batson. He developed the empathy-altruism hypothesis, which states that only empathy leads to true self-less helping behaviour (i.e., altruism). In terms of intergroup relations this is also interesting as social psychological research shows that indeed when engaging in helping behaviour people discriminate in favour of their in-group, so they would help more members of the same race, religion, or gender, compared to helping someone from an out-group. Thus increasing empathy for out-group members could also reduce discrimination and increase helping behaviour. In their study Batson manipulated empathy by creating similarity. In particular, the person who was later going to be in need was either described as similar or dissimilar (in personality traits) to the participant. Basically, participants read personality descriptions of the person (for example the other participant is caring, outgoing, likes sports), but this description was manipulated as it was made either similar to the subjects own interests and personality traits or dissimilar to them. On the testing day – after filling out demographic and personality questionnaires – the participants were brought to the laboratory and then read the personality description of Elaine. Elaine was a confederate of the experimenter, but the subjects were told that she was a real participant as well. For half of the subjects Elaine was apparently similar in interests and personality to themselves and for half she was not. Then the participants were brought into a different room in which they watched Elaine reading out words. Every time Elaine made a mistake she would receive electric shocks (not for real). Elaine then said to the experimenter that she was very much in pain from the electric shocks and that she could really tolerate it anymore. The experimenter then offered the participant to take the shocks themselves, relieving Elaine. A further manipulation was made here; for another half of the participants the experimenter mentioned that if they did not want to take the shocks themselves they would still though have to stay in the room and watch Elaine getting more (difficult to escape situation). In the other group participants were told that if they did not take the shocks themselves then Elaine would get them but that the participants could leave the room and did not have to watch her (easy to escape situation). The results were that more people helped when it was difficult to escape. The reason is obvious; the participants felt uneasy having to watch Elaine receive the shocks, so if they had to continue watching her, and were not allowed to go away then they would rather help and take the shocks themselves. However, and this is the interesting part of the finding, in the group where empathy was increased (by describing Elaine as similar to the participant) the subjects also helped equally when it was easy to escape. This suggests that empathy increased help.

One interesting observation in this study is that empathy was induced by making Elaine similar to the participant. Indeed, evolutionary psychology suggests that helping one's own kin (e.g., those similar to oneself) might have been a trait developed through natural selection. Historically if there were individuals that possessed genes (gene combinations) that made them more likely to help their own relatives (those with whom their shared genes) then this would lead to a population of genes having an evolutionary benefit, thus being represented more in the following generation. Specifically, helping one's own kin has an evolutionary benefit as it promotes one's own genes to be represented more in following generations. Thus, it can be suggested that helping one's own family and relatives is a trait which is predisposed. Regarding prejudice and intergroup relations; increasing empathy might be a way to reduce prejudice and discrimination.

In the remainder of this chapter I will discuss a further very important issue; previously I have discussed social psychology experiments on prejudice. But how do we measure prejudice? How do we know if someone is racist? What are the tools available for social psychologist to assess if someone is prejudiced? And, are they good tools? When assessing prejudice, researchers often distinguish between explicit and implicit biases, which is partly referring to the measurement method, but also partly referring to independent theoretical constructs of prejudice. First, I will discuss explicit prejudice, as it's easier to explain. Explicit prejudice is measured with simple self-report questionnaires. There are a variety of questionnaires available. In one questionnaire for example, which is called the feeling thermometer, participants indicate how warm they feel towards various groups (e.g., black people, white people, Christians, Muslims). Participants do this by ticking a box on a 10 point scale which ranges from 0 degree (very cold) to 100 degree (very warm).

In a different questionnaire participants see different adjectives that are opposite (for example positive versus negative; fear versus calm etc.). Here participants are asked to tick a box on a 7-point scale ranging for example from 0 (calm) to 10 (fear). This way of assessing prejudice is, as the name implies, explicit; one is simply *asked* about their feelings towards different groups. However, as discussed before, in western democratic countries, and within mostly student participant populations, who is going to admit they feel different to other people based on their race or religion? This is a question that is often asked. Indeed, the problem with explicit measures is the fact that participants might not report their true feelings (however it is important to note that the questionnaires are anonymous and experimental ethics prohibits researchers from disclosing any individual responses to anyone outside the research team. In addition, responses are also coded with numbers and not saved together with the participant's name, but only with their participant number.

I presume that by now the reader might think that if I give such questionnaires to western undergraduate psychology students no one would indicate they feel better about their own group. Well, that's not correct. Even though most participants report that they do not have different feelings towards in-group and out-group members, in many different experiments (at least 50 different studies) at different universities, with different populations, with different experimenters, I have never found explicit prejudice to be completely zero. Indeed, I never had a sample of participants where every single participant said that they felt the same towards black and white people. Indeed, explicit prejudice questionnaires have shown to have high reliability and validity, which are indicators of the 'goodness' of a test. For instance the explicit prejudice measure correlates with outward discriminatory behaviour. If someone for example has a very high score, e.g., suggesting say that they felt much more positive towards white people than black people; behavioural observations are also more likely to show more hostile towards black people.

Now I will turn to the second concept, that of implicit biases. Implicit biases, as I mentioned before are partly distinct from explicit because of their measurement method, but also because they might represent a different aspect of intergroup prejudice. Implicit biases are measured with indirect methods. Indirect methods, compared to explicit or direct methods, as its name implies, measure a concept with less overt methods. For instance if one is interested if someone was frightened by a horror movie, other than merely asking the person, one could also observe their behaviour (i.e., do they close their eyes, body language) and one could also use physiological methods, such as recording the persons heart rate. In measuring prejudice, there are also many studies that use physiological research methods, such as fMRI, EEG, skin conductance and heart rate, and most find differences in those measures when people view in-versus out-group stimuli. I will discuss such physiological measures more in Chap. 3. In this chapter I will conclude by describing the most commonly used behavioural method to investigate racial biases, the Implicit Association Test (or the IAT).

Tony Greenwald developed the IAT in 1998, and to date it is the most widely used test for implicit biases. In fact, according to the Google Scholar search engine, his original paper on the IAT has now been cited 6487 times by other researchers.

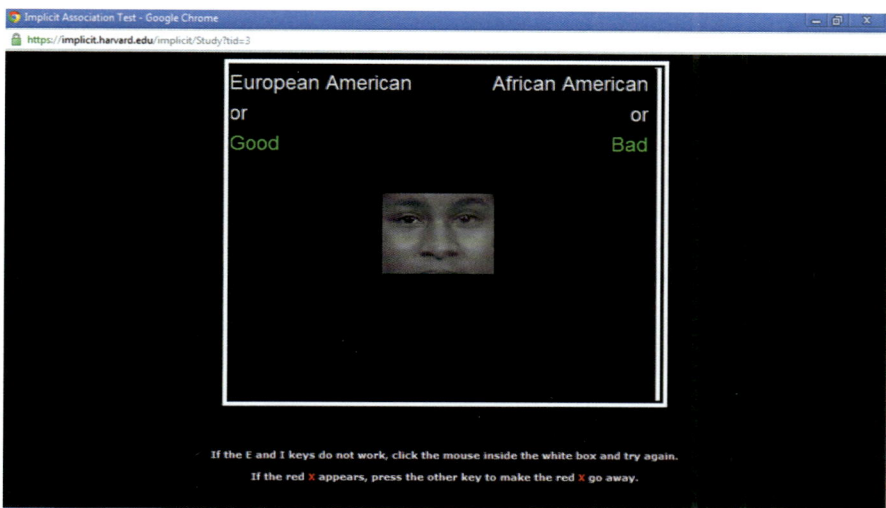

Fig. 2.1 Racial Implicit Association Test (IAT). Retrieved from http://implicit.harvard.edu/implicit/takeatest.html (Reprint permission obtained from emily@projectimplicit.net on 05.07.2016)

The IAT is a response time based computer task. I will describe it here, but I also want to encourage the reader to complete it themselves, as a free version is available online under; http://implicit.harvard.edu/implicit/takeatest.html. There are different IATs, for example one can test their racial biases, religious biases, a Fat-Thin IAT, a gender and science IAT, and many more. I will describe the test in detail here for several reasons; first, as most researchers, including myself, use this test, anyone interested in the social psychology and neuroscience of intergroup relations should know about it. And secondly, because in this book in the final chapter I will try to answer the question "Prejudice, can we cure it?", so people should know, how I measured biases. Participants start the test and they first see the instructions which tell them to place both their index finders on the "e" and the "i" keys of the keyboard during the test. They will then see a screen similar to the one shown in Fig. 2.1.

Figure 2.1 shows only one particular block of the IAT (the bias congruent block if participants are Caucasian and have a bias). What one can see is that there are two categories in the left and right corner of the screen, and there is a picture of a black man in the middle of the screen. In each block the categories in the corners remain constant, but the images in the middle change. Also in the middle of the screen there are not just pictures but on some trials there are also words (such as sunshine, warm, danger etc., which are of positive or negative nature). The task for the participant is to sort these items (pictures and words) in the middle to the correct left or right corner (by using the e and i key of the keyboard). As shown in Fig. 2.1, the picture in the middle is the picture of a black man so the correct answer would be to press the 'i' key to sort the image to the right side. On the next trial the subject might see the word "sunshine" in the middle, and then the correct response would be to press the "e" key to sort this item to the left corner. Now the block shown in Fig. 2.1 is a 'congruent' block for someone who has a bias. In this block they are sorting all pictures of white people and positive words to the same corner (here the left corner) and they are sorting all pictures of black people and all negative words to the right corner. However, this is only one block. In another block of the test, the categories in the left and right corner will reverse. In that block the participants would have to sort all white faces and negative words to one corner (say the left corner) and all pictures of black people and positive words to the right corner (this would be the bias incongruent block). Also note that participants are asked to do this as quickly as possible. The computer then records the response times for each block. What numerous researchers now find – and this is what you might have found as a result for yourself if you completed the test online – that white participants have an increased response time for the incongruent block (where they sort "black/good" and "white/bad") compared to the congruent block (where they sort "black/bad" and "white/good"). Thus the association of black and bad words and white and good words for them is easier (i.e., faster response time), compared to the association of black and good and white and bad. This is also why the test is called an "association" test, as it is looking at which concepts a person has closely associated. The striking finding with the test is that the vast majority of people, regardless of the score on explicit prejudice measures, have a non-zero value on this test. Indeed, most people have an implicit racial bias on the IAT.

So what? We are all racist, even though we don't know, or don't want to be? No. It is more complicated than that. Indeed, researchers found that American radicals who were committed to racial justice, still, showed an IAT effect. Moreover, researchers find is that when black people complete the IAT they *also* have a bias; the same bias as white people, so black people also find it easier to associate black with bad and white with good. Even though most people – when being sloppy – report IAT as a test of implicit 'prejudice', or indeed racism, no underlying theory supports this conclusion completely. So we have to look into *what* the IAT is measuring and if indeed it can be related to prejudice. What the IAT *is* measuring is associations that people make, but are these associations based on prejudice? In other words, does the IAT measure (implicit) prejudice? Furthermore, racial biases (e.g., IAT scores) are found to often show a low correlation with explicit prejudice scores. This indicates that individuals with sincere believe in tolerance and equality can however still show implicit biases. This would also suggest that some individuals might have attitudes that they are not aware of. Is that possible? This is a complex question.

From the research we know that the IAT in fact cannot measure in-group out-group biases, as if this was the case black people would not also have an IAT effect or indeed would have an IAT effect in the opposite direction. In Chaps. 3 and 4 I will present neuroscience evidence that the IAT correlates with many factors involving emotional arousal responses, and thus seems to be associated with more emotional aspects. It is often suggested that the IAT must measure what one has learned or what was presented in the media, for instance a less positive image of black people. This would then refer to the stereotype people have, but not the stereotypic knowledge but rather the emotion associated with this. For instance it was found that black people were often portrayed in films in negative roles (the gangster, or the criminal rather than the police officer or the lawyer). This might have already partly changed as part of equality and diversity progress but it might still create or support stereotypical images with are associated with certain emotions. This then strictly speaking is different to prejudice – as defined before as the attitude someone has to a group. Because the emotions that are generally associated with a certain group does not necessarily have to be the emotion that *one* holds too. But again this is more complicated than that.

If this was true, then there are two problems that arise; (a) why do people have different scores (as all have similar idea of the stereotype) and (b) why does the IAT correlate with real life behaviour? Individuals greatly vary on their final IAT score, and previous research has indicated that this score predicts certain real life behaviours. And indeed, this is what makes the IAT robust, as researchers find that those individuals who have a high IAT score also behave differently towards people of different races. For example in one large study researchers found that the IAT correlated with the chair distance participants used for a black but not white experimenter. Specifically, white people with a higher IAT score also sat further away from a black person. A second behaviour that the IAT predicted was the number of positive versus negative words used in an interaction with people of different races. These behaviours are often termed, subtle discrimination, indicating that explicit prejudice predicts overt discrimination whilst implicit biases predict subtle behaviour differences. I have now illustrated that behavioural experimental work on prejudice has made progress but further research is still needed.

Here I would like to add another paragraph about the use of the IAT. Sometimes there might be misunderstandings, and indeed people might want to use this as "a test of racism", or a test to show that "you are biased", even though you don't know it. This is not possible. Firstly, one should note that up to 90 % people show a bias on the IAT. And this refers to everyone, men, women, black, white, Christian, Muslim. It is thus not possible to say that someone in particular has a bias but others don't. As I discussed before, humans have the tendency to form groups and to perceive others as out-group and in-group. Secondly, the IAT is a research tool, and cannot be used out of this context. Indeed, a recent large meta-analysis found that the IAT was not a good predictor of any behaviour outside the lab, even though, as discussed above some studies do suggest certain behaviour links. Thus, if the reader should determine that they show a bias on the IAT, or indeed if someone was tells you "You are implicitly biased." they could well reply;" Well, so are you."

But let me finish with a final 'amazing' study. A research team led by Prof. Neha Mahajan (University of Santos) showed that chimpanzees also have an IAT score. How can monkeys sit at the computer and do an IAT? The researchers created a non-verbal version of the IAT, in which they showed the aps pictures of in-group monkeys paired with spiders (which they fear), or fruit (which they like). Monkeys looked longer at the pictures that show in-group with spiders compared to outgroup, and longer at those that compared fruit with outgroup compared to in-group. Amazing!

Open Questions Chapter 2

- Have you faced prejudice?
- How do you determine something is prejudice, and something is not prejudice?
- Do you think people too easily assume someone is being prejudiced or should the laws be even stricter?
- Do you think everyone is prejudiced, or are there people who are not?

References

Allport, G. W. (1956). *Prejudice in modern perspective*. Johannesburg: SA Institute of Race Relations.

Berreby, D. (2005). *Us and them; Understanding your tribal mind*. New York: Little, Brown & Company.

Diamond, J. (2012). *The World until yesterday*. New York: Penguin Press.

Follett, K. (2012). *Winter of the World*. London: Penguin Press.

Gracia, J. L. A. (2004). Three sites for racism. In M. P. Levine & T. Patki (Eds.), *Racism in mind*. New York: Cornell University Press.

Greene, J. (2003). From neural 'is' to moral 'ought': What are the moral implications of neuroscientific moral psychology? *Nature Reviews Neuroscience, 4*(10), 846–850.

Guhman, P. (2015). Reaching out to the untouchables. *Psychologist, 28*, 564–570.

Kurzban, R., Tooby, J., & Cosmides, L. (2001). Can race be erased? *PNAS, 98*, 15387–15392.

Leslie, S. J. (2015). The original sin of cognition: Fear, prejudice and generalisation. *Journal of Philosophy* (forthcoming).

Morrison, T. (2006). *Song of Solomon*. London: Vintage.

Picoult, J. (2013). *The storyteller*. London: Atria Books.

Preston, A. (2014). *In love and war*. London: Faber & Faber.

Ronson, R. (2001). *Them*. Oxford: Picador.

Sherif, M., & Sherif, C. W. (1969). Ingroup and intergroup relations: Experimental analysis. In *Social psychology* (pp. 221–266). New York: Harper & Row.

Tajfel, H., & Turner, J. C. (1979). An integrative theory of intergroup conflict. *The Social Psychology of Intergroup Relations, 33*(47), 74.

Turner, J. C., Hogg, M. A., Oakes, P. J., Reicher, S. D., & Wetherell, M. S. (1987). *Rediscovering the social group: A self-categorization theory*. Oxford: Basil Blackwell.

Walsh, M. (2010). *Gypsy Boy*. London: Hodder & Stoughton.

Chapter 3
The Neuroscience of Prejudice

Brain research will be the central topic in this chapter. Firstly, I will introduce the basic nervous system and its anatomy, secondly I will describe neuroscience research methods used to investigate brain functions, and then I will review research that has investigated the brain and intergroup relations. I appreciate that some readers already have an insight into neuroscience and brain anatomy, but I will revise elementary neuroscience facts here before proceeding. It is often said that neuroscience is *the* popular, new, and promising field of the future. For example, in his book on neuroscience, Restak (2004) offers the promising prospect that neuroscience has now allowed us to "study the brain in real-time, when we are thinking, taking an intelligence test, practicing craft, experiencing an emotion or making a decision." Even though neuroscience does indeed provide new insights, there are however also limitations to those possibilities. The brain is the most complex structure in the universe, so it's difficult to know where to start. We don't understand it fully yet, and enormous sums of money are devoted to further research it. The human brain mapping – the BRAIN Initiative – project was funded with billions, precisely with $300 Million per year for ten years – starting 2013 with its aim is to map all neuronal connections in the human brain. Figure 3.1 shows one neuron.

The blue ball with the spikes (it is only blue in this image, just in case someone was wondering) shows one neuron or nerve cell and it is the unit of the central nervous system. It has been found that the human brain has 100 billion neurons. But, that's not quite correct; I recently read an article stating that the latest research has shown that it was only 86 billion, on average. And these 86 billion neurons have on average 100 Trillion connections. So the 'Brain Initiative' is certainly ambitious. Investigations into the brain require a large interdisciplinary field, ranging from biologists, to researchers in physics. Describing the brain and its functions is again a challenge, and usually books, even textbooks, can only provide a very simplified version of the real picture. This means that one reads about one molecule, e.g., oxytocin, being involved in feelings of love. Or one reads about how serotonin is involved in depression, which might lead to the conclusion that this is the whole story; 'Serotonin is low, you are sad'. But as one goes deeper and deeper into this

© Springer International Publishing Switzerland 2016
S. Terbeck, *The Social Neuroscience of Intergroup Relations: Prejudice, can we cure it?*, DOI 10.1007/978-3-319-46338-4_3

Fig. 3.1 One neuron (Retrieve from http://www. sciencedirect.com/science/ article/pii/ S0956566315300610. Reprint permission code: 3902531225766)

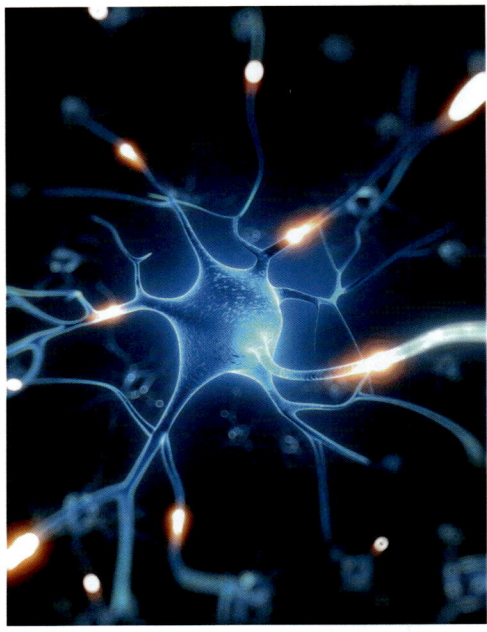

field the picture becomes a lot more complex. For example I recently wrote a paper about the effect of glutamate in anxiety. But, in fact, if you want to have the more complex picture, I wrote a paper about how one of many glutamate receptors (the metabotropic glutamate receptor number five), is involved in mediating secondary biochemical processes involved in some forms of anxiety in an average person. This might make clear that any one book, or any one person, or any one discipline cannot give the full picture of neuroscience. For example I presume that a biochemical researcher could write a whole book solely on the topic of *one* glutamate receptor. So I will just give a simple overview of neuroscience, one that might be sufficient to understand the research that then looks into the brain and 'prejudice'.

We know that one key function of the neurons in the brain is to transfer signals from one to another. Neurons (nerve cells) conduct electrical signalling and chemical synaptic communication. This is responsible for the transfer of signals between neurons (communication) within the brain. Neurons have a soma (cell body), dendrites (to receive signals from other neurons) and one axon (to transfer the signal to another neuron). The gap between two neurons is called a synapse, where a chemical transduction of the electrical signal is performed (and where drugs can exert their effects). When 'inactive' a nerve cell has a so called resting membrane potential, serving as a medium to conduct electromechanical signalling. The fluid inside (intracellular) and outside (extracellular) the neuron contains positively and negatively charged molecules (ions). At rest, the intracellular fluid has more negatively charged ions (the membrane potential is around −70 mV). This results from the distribution of sodium, potassium and chloride ions. The membrane's permeability to potassium is greater than its permeability to sodium resulting in more extracellular sodium on the outside of the cell. The membrane at rest is therefore polarised. When a neuron is stimulated by an electrical or chemical signal it is said to be depolarised. The membrane potential changes from −70 mV to +30 mV and an action potential or a spike is generated. During this phase of action potential generation, voltage gated channels open allowing an influx of sodium ions to rush into the neuron. The spike arises at the trigger zone (or axon hillock) and propagates down the axon. As Na+ ions flow in during depolarisation at one membrane location, voltage-gated Na+ channels at adjacent locations open too. The wave of depolarisation propagates along. Repolarisation follows as the electrical signal travels down the length of the axon of the neuron. This is known as the generation and propagation of an action potential (see left panel in the Figure below). The signal is transferred between two neurons by chemical signalling at the synapse. The propagated electrical signal is conducted down the axon (middle panel of the Figure) until it reaches the axon terminal where it triggers synaptic vesicles, which contain neurotransmitter to release it into the synaptic cleft – the space between the axon terminal of the presynaptic neuron and the postsynaptic neuron or muscle. The neurotransmitter drifts across the synapse to the membrane of the postsynaptic neuron upon which there are receptors to which it binds (right panel of the Figure below). Once the postsynaptic neuron's excitation is complete, it becomes a presynaptic neuron and transmits information as described above. Such communication between neurons and assemblies of neurons results in information processing in the form of perception and action allowing the brain as a whole to react to its environment.

The study of neuroscience and brain imaging has informed us of those neurons in specific regions of the brain that are activated upon 'seeing' stimuli such as a smiling face, or a frowning one. Later in the chapter, I will describe which methods can be used to determine this. The brain has many structurally and functionally defined regions, which are not obvious or detailed when one looks at an image of its lobes (see Fig. 3.2.). But, the brain is made up of many sulci (valleys) and gyri (spurs), which house all its many functional units. The brain can be divided anatomically into three main parts; Cerebrum, Cerebellum, and brain stem. The cerebrum is divided into two hemispheres– one on the right and the other on the left. The hemispheres are connected to each other via the corpus callosum, which is made up of white matter or myelinated axons of neurons. Cerebellar hemispheres are connected to the midbrain via cerebellar peduncles. The cerebral cortex can be divided into four lobes: frontal, parietal, temporal, and occipital. The grey matter of the cerebral cortex is made up of neuronal cell bodies and these are on the outside. The white matter tracts are made up of the insulated axons of the cell bodies of neurons and are on the inside. They include the commissural, projection and association fibres. The commissural fibres connect the two hemispheres via the corpus callosum. Projection fibres project to and from spinal cord and sub-cortical structures. Association fibres link neurons within the same hemisphere.

It is important to understand though that even though regions are anatomically defined they do not correspond to single functions. Even though the brain has locations which might predominantly be involved in certain psychological concepts,

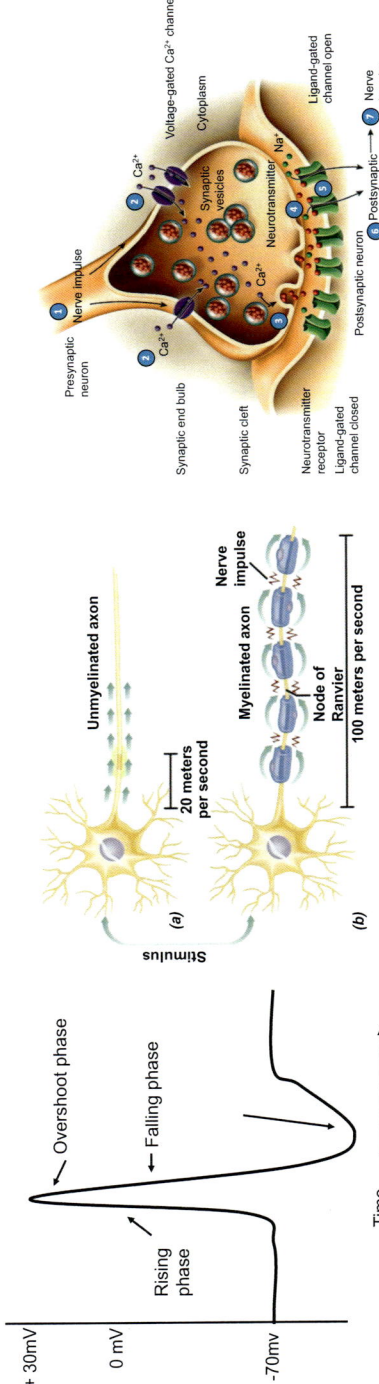

Fig. 3.2 Synaptic transmission (Copyright ©John Wiley and Sons 14th edition of the book: Principles of Anatomy and Physiology by Gerald J. Tortora and Bryan H. Derrickson (copyright received on 15.07.2016))

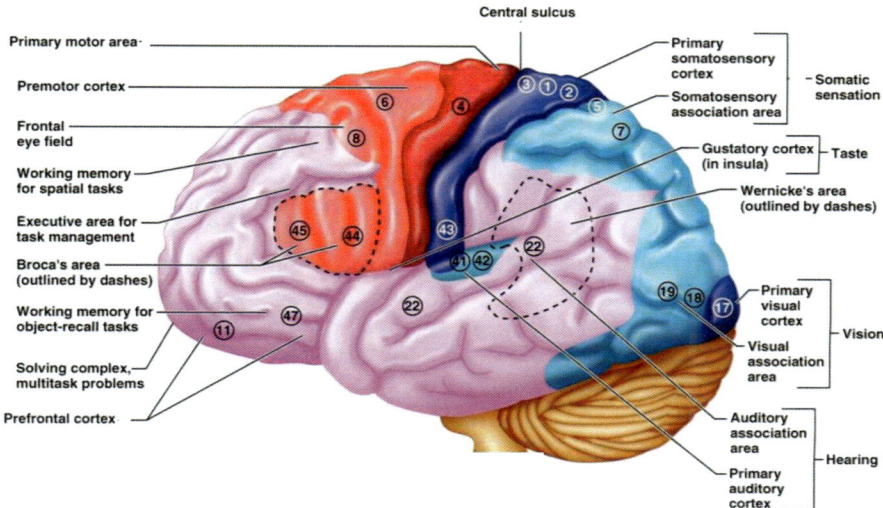

Fig. 3.3 Brain anatomy (Retrieved from http://www.turbatour.com/page/185, permission received on 15.07)

which I will briefly discuss in the following, there is no single function that can be located to one – and only one – area. Also, there is no one area that is only involved in one – and only one – function (See Fig. 3.3.). The lobes of the brain were further sub-divided according to neuronal cytoarchitecture (the study of the brain's cells under a microscope) by German anatomist Korbinian Brodmann in 1909 (areas numbering from 1 to 52).

For example, it was found that the hippocampus is strongly involved in forming new memories. The hypothalamus is thought to be involved in regulating physiological state of the body, the thalamus in selecting and filtering sensory information, and the cerebellum in controlling posture and movement. However, even though these regions might be predominantly associated with a certain function, the complete functional network is always larger, involving more brain regions. Compared to an animal brain, the human brains' frontal cortex is more developed and has been associated with higher order functions; that is, functions related to morality, self-control, inhibition of impulses etc. Indeed, the prefrontal cortex was suggested to be involved in controlling the processing of emotional responses. For example, if now, while you are reading this, someone behind you said: "Boo!" you might have an instant fear response, but might also be angry. However, subsequently you will regain control over this situation and indeed not report to 'fight or flight'. Research has shown that damage to the prefrontal cortex, amongst others, impairs emotion control, and an inability to inhibit aggression. With regards to aggression, researchers found that when imagining an aggressive, compared to a non-aggressive response, reduced activity in the prefrontal cortex was found. To repeat, after reading this, one might still think that it would be easy to believe that there was after all *one* area and *one* function – for example frontal cortex = morality. However, that is not true. Indeed, to start with there aren't as many areas as things we do (crying, eating, smiling, running, lying, worrying, reading …). Thus, when discussing the prefrontal cortex, and when I write that this area was associated with self-control and morality, what I mean is that research has shown that within a network of different areas, neurons in the prefrontal cortex are also active when a person is processing information related to morality, as compared to when they are processing something else.

Inside the brain – in the inner layers – we can determine amongst other structures the so called limbic system (but areas of the limbic system also extend to the cortex). The most prominent neural areas of the limbic system are the cingulate gyrus, the hippocampus, and the amygdala. In 1961, Downer surgically removed one amygdala (normally one has 2 amygdalae, one on each side) of an experimental animal. If the monkey then interacted with the world via the eye that was connected to the functioning amygdala, it was found to behave normally, also showing fear and aggression when provoked. However when the eye was covered, so that the surroundings were viewed with the eye related to the non-functioning amygdala the monkey was quiet and could be easily handled; indeed the animal was unable to perceive threat situations and respond emotionally towards them.

The eminent neuroscientist LeDoux also made significant contributions investigating the role of the amygdala in fear learning and fear memory. This initial research supports the strong role of the amygdala in threat processing and emotional responding. In addition, animal research has also shown that electrical stimulation of the amygdala induces fear responses in the animal. Research has suggested that in humans, even subliminal stimuli can activate the amygdala. Subliminal stimuli are outside of conscious awareness, this means they are displayed for such short durations that participants report that they have not seen anything. The Internet provides us with the opportunity to view and study the brain in 3D, which is an impressive tool for someone who might be interested in studying anatomical regions of the brain further. Also there are free tablet apps, for example a 3D anatomical brain atlas, that give an impressive view of the brain and its structures. The key fact to remember might be that neurons within particular brain regions are more or less active at times, and that some of this activity in a network of brain regions can be associated with certain human functions.

Students often learn very briefly about many neuroscience methods, fMRI, TMS, PET, EEG, etc. etc. Here, I will mostly describe one method, one that is most often used to investigate which areas of the brain show increased activity under certain psychological task, the method of functional Magnetic Resonance Imaging (fMRI). The fMRI method is very complex. Now this sounds like 'everything is complex'; first I said the brain was complex, anatomical regions were complex, biochemistry was complex, and now fMRI is complex too. Indeed, that is one key message; even measuring the brain – which itself is complex – is also complex. The validity of the experiment depends on the task used during the fMRI scan, it depends also on the way the brain is examined, and it depends on the way the data is analysed – thus it *is* complex. Even though an fMRI expert will have expertise and experience in analysing the experimental data, I presume they would still not be able to build an fMRI machine, or to understand fully all the processes involved, which might need an expert in mechanics, an expert in physics etc. Indeed, there are also whole books just on fMRI.

The MRI machine, which looks like a tube, (the participant is lying inside the tube) has a very strong electro-magnet. The participant is inside this huge magnet. Thus they could obviously not take any keys or coins etc., with them (anything magnetic) as the strength of the magnet (for example 3 Tesla) are 50,000 times greater than the Earth's magnetic field. Atoms have magnetic nuclei and their orientation is usually random, but inside the fMRI and its huge magnetic field the arrangement of the orientation of the atomic nuclei can be changed. FMRI measures the magnetic signal from hydrogen nuclei in water ($H2O$), and changes in the magnetic properties of the haemoglobin molecule. When a certain brain area is active, then its neurons need more oxygen. Blood in this area has more deoxygenated haemoglobin molecules (because the neurons are using it), and deoxygenated haemoglobin has different magnetic properties to oxygenated haemoglobin. The basis of the fMRI measure is the change in magnetic properties of blood due to the shift in ratio of oxygenated to deoxygenated haemoglobin Because this correlates with the change in neuronal activity we can estimate how a particular psychological task has altered activity in different parts of the brain.

FMRI is not yet sufficiently advanced to be able to detect any brain activity very easily. For example, it is not possible to put a person into an fMRI scanner, let them watch a film, and then from watching the brain activity 'read back' to see what the film was about. If it was that easy, one could then literally read a person. For example, we could put them into an fMRI scanner, ask them about things, and if they were to lie we would know that they had lied. Or we could put someone into an fMRI scanner and ask "are you prejudiced?", and they would not even have to say – we could just look at the brain activity and we could know. Obviously, this is not possible, because if it was possible, we would be doing this all the time. But there are some studies that go in this direction – for example we *are* able to look at the brain activity and without knowing, we can see from the pattern of activity if for example someone is seeing a face or an object (though that does not hold for all objects).

Have you seen the cool new tool that reads your mind? Precisely, it measures brain activity and then you can merely think about moving a ball which appears on a computer screen and then it moves (up and down, to the left and right etc.). Basically, you can move an object just by thinking about it. You don't have to speak in order for it to move you can just think it and it will happen. Honestly though, I tried that kit, and you have to think the thought (moving the ball) *a lot*. You first have to put that kit on your head and record the baseline, which is the activity when you think nothing in particular. Then you get asked to think "Move up". On the screen you see a ball and you get prompted to think "Move up". So I was then thinking "Move up", – then a ball moves up on the screen, and the kit saves that activity. Now I can see the ball on the screen again and the logic is that it has saved my brain activity pattern for the thought "Move up" and every time I am thinking this again the ball really does move up. So I was looking at the ball – not moving – and thinking "Move up". Nothing. I was thinking again "Move up" "Move up". Nothing. "Move up" "Move up" Move up!" "Dammit MOVE UP". And then it did it. I gave up trying to move it to left and right. Indeed, it is impressive that we can do that, but you can see it's not optimal yet.

One problem with fMRI is that we need to average the activity over a number of trials that contain the same information, for example in order to see which area is active when someone sees a face one has to show a face more than once. Also, the problem is that the brain is never – ever – inactive. If you are thinking nothing, or sleeping (and not even dreaming), there is always a lot of brain activity. Because the brain of course controls all our functions, such as hunger, thirst, our body movements. Everything. Thus, in a typical fMRI study one uses very simple stimuli, such as, participants' seeing a face, or seeing a chair, or seeing a hand etc. And then it needs a control condition in which they also see something that is similar but varies according to the aspect of investigation. For example they see a smiling face and an angry face, and a neutral face. Usually what the researchers then find is different activity for the emotional faces (the smiling and the angry) compared to the neutral face. And as I said before, such a scan would take about half an hour, because they would have to see a lot of smiling, angry, and neutral faces. Thus the fMRI is always a contrast between two conditions. The first is between the activity and the baseline, which is when you see a face (any face) compared to when you see nothing. And the second would be the contrast between seeing the emotional and the neutral face. Thus the most common procedure is to see a simple stimulus many times, within different conditions.

What then about assessing more complex concepts, such as morality? In such an experimental setup usually participants read something related to morality (e.g., he killed her), compared to something neutral (e.g., he knows her), compared to something emotional but not moral (e.g., he kissed her). And again this needs multiple sentences in which some contain words related to morality (killing, lying) and others don't. However, it becomes complicated to choose comparisons here, because "he kissed her" is also positively emotional, and not just non-moral, and indeed for some people this might be also moral. Thus, this is another reason why many trials are needed. It is the same problem with showing angry and happy faces. Which genders do the faces have? Is the visual contrast of the faces the same? Are they the same age? There are thus multiple issues to consider when designing an fMRI experiment. In terms of investigating fMRI and prejudice, the commonest way is to show many faces of different races, or ages or with characteristics associated with different religions. So for example participants could be shown black and white faces that were rated before so that they differ in nothing else (gender, age, attractiveness etc.) other than the racial aspect. This would allow investigating if there are differences in brain activity pattern between the faces based on the group component.

After having reviewed how an fMRI design might look like, I want to proceed to describe very briefly how fMRI data are usually analysed. Now why would anyone reading this book be interested in how fMRI data were analysed? I presume that if someone really wanted to learn to analyse fMRI data, they would not read my book for that reason, but instead buy a specialised book. I will only briefly describe fMRI data analysis in order to demonstrate that this is not a simple process. Indeed, the first time I ever analysed a set of fMRI data I found that when participants viewed faces in the fMRI scanner they had brain activity outside of their brain. Well, if that was true, I'd have made the biggest discovery in the history of science; you see a face and your neurons move out of your brain. Then I thought the simple explanation is that I did the analysis wrong. The fMRI data comes in 'messy', which means there is a lot of random – not stimulus related activity – which needs to be excluded. First though, before this, the data needs to be filtered, with high/low pass filters of certain characteristics and bandwidths. And also (this is where I was getting it wrong the first try) it needed to be matched to the individual brain of each participant. Thus during the scan, besides the task, the experimenter would also record an anatomical scan of the participants brain, to which the data is then fitted. A further step is that the individual scans need to be averaged. Basically, the whole process involves multiple complex steps, using many calculations, filtering, extractions, fittings etc. etc. before anything can be observed. And indeed, if one was not an expert, analysing such data all the time, any difference between any of those mathematical processes can lead to very different results. Thus in order to conduct fMRI research it is essential to have a team of experts involved at all stages. Here is an extract from our methods section of an fMRI study; *all imaging data were obtained using a 3T Siemens Tim Trio system. Functional imaging consisted of T2*-weighted echoplanar image (EPI) slices [repetition time (TR)=2000 ms, echo time (TE) = 28 ms, matrix=192×192], with 3.5-mm slice thickness. The first two EPI volumes in each session were discarded to avoid T1 equilibration effects...*

As discussed above, besides fMRI research, there are of course a number of other neuroscience approaches, and experiments using further physiological methods to investigate psychological functions, for example, EEG, heart rate, skin conductance etc. Below, when discussing neuroscience research of intergroup relations, mostly fMRI studies will be mentioned, because most studies in the field have used fMRI, but other relevant neuroscience methods will also be mentioned.

3.1 Neuroscience Research of Intergroup Relations

Cikara and Van Bavel (2014) stated in their review article that the brain is truly social, specialised for group living. Liebermann (2013) suggested that we are wired (or born) to be social. Indeed, just as we need food and drink, we also need social interaction in order to function. Humans are social animals. He also suggested that reaching out and interacting with others makes humans the most successful species on earth. Indeed, he argues that for many years we thought that maybe the smartest humans were those with the highest intellectual abilities, but maybe it was that that smartest were the ones with the highest social abilities. Furthermore, the anthropologist Robin Dunbar suggested that the size of the neocortex grew larger in evolution so that primates could live together in larger groups and be more socially active. He used correlations to investigate factors that are associated brain size development, such as individual innovation and social learning, but he found that group size was the strongest predictor for neocortex size. Indeed, even for chimpanzees, living in larger groups requires an enhanced set of social skills, such as observing and choosing suitable companions and forming coalitions in order to succeed socially.

The study of social psychology and neuroscience combined opened a new field of research entitled social neuroscience. The study that is often mentioned first in review articles on neuroscience of intergroup relations was conducted by Phelps et al. (2000). In this study the participants – all Caucasian – passively viewed black and white faces during the fMRI scan. Then, outside the scanner, participants completed a racial IAT and further explicit prejudice questionnaires. Phelps et al. (2000) found that the amygdala activity correlated with the IAT, meaning the higher the implicit bias the higher the amygdala activity to black versus white faces. As I briefly mentioned before, the amygdala is a brain area involved in the processing of emotions. Indeed, numerous studies have shown that the amygdala is active when participants view emotionally arousing stimuli. A simple and common interpretation of the results in Phelps study is: "The prejudiced participants are frightened of outgroup faces". Now, this interpretation is wrong on two counts. Firstly, as discussed in Chap. 2, the IAT does not necessarily measure prejudice (at least it is more complicated than that), and secondly, the amygdala activity does not only reflect fear responses, but *any* emotional reaction (e.g., aggression, happiness, and sadness).

After all, participants with high IAT scores could be really happy to see faces from another race. Whilst this might be unlikely, the interpretation however cannot be excluded on the basis of Phelp's results. Also it could be that emotions such as aggression, sadness, or a more complex combination of all emotions are involved in prejudice. However, what this result suggests is that individuals who have a high IAT, process out-group faces (emotionally) differently. That is interesting, and also suggests that faces are perceived differently – or evoke different emotions – and this difference is based on the race of the face. However, this was only one study; but there were many to follow. As stated before, Phelps et al. (2000) did not actually find a difference in the amygdala activity between black and white faces, but only that the amygdala activity was correlated with the IAT. However, later studies found just that; that the amygdala was more active when viewing black, as compared to white faces. One such study was conducted by Cunningham et al. (2004). Compared to Phelps et al. (2000) however they showed the faces for different time durations. For example, on some trials they showed the faces for 2 s (which is quite long), and on other trials they showed them for only 300–500 ms (the faces flashed on the screen and then disappear again). The authors found that when the faces were shown for a short duration there was a difference in amygdala activity between black and white faces (participants were Caucasian). When the researchers showed the faces for longer, then the amygdala activity disappeared but they found activity in the ventro-medial prefrontal cortex (vMPC). The authors suggested that the participants, when they see the faces for longer, consciously controlled for their emotional bias, as the vMPC was previously found to be active when participants were engaged in cognitive control processes. After this, many more fMRI studies were conducted investigating the neuronal biases of intergroup relations. Lieberman (2003) also suggested that self-control (via vMPC) not only serves individual aims (such as exerting self-control by not eating fish and chips while on a diet) but also social aims. For instance self-control can lead to reducing the focus on oneself and thus also helping the group, and it also helps individuals to stick to social norms and comply with society's rules.

Researchers used this technique not only to investigate racial prejudice but also prejudices related to age, gender, obesity, etc. And there were further key findings identified; two other brain regions that seem to be involved in the processing of in-group/out-group bias- the insula and the fusiform gyrus. The insula is a brain region that was found to be associated with the processing of disgust. Specifically, parts of the insula were found to be active when viewing disgusting images (for example of vomit) versus neutral images. Some neuroscience studies have now found that the emotion of disgust might also be relevant in social interaction/perception as well as in moral decisions. In everyday language disgust and morality are indeed sometimes coupled; for example if someone does something we might find morally wrong people might say that they find that act "disgusting". For example, if a child steals from a friend, their mother might say "Your behaviour is disgusting". Indeed, in a large experimental study Jonathan Haidt found that participants judged disgusting actions as morally wrong (for instance cleaning the toilet with the American flag). Within intergroup relations, researchers found that that the perception of some social groups (e.g., overweight people, homeless people) correlated to activity in the insula, which was thus, interpreted as disgust related emotions being related to social perceptions. Some researchers have suggested that different social groups might elicit different emotional reposes within the average person.

The stereotype content model combines participants rating of groups on the dimensions of warmth and competence. For example, viewing the elderly or disabled was found to elicit feelings of high warmth and low competence, thus, leading to feelings of pity. On the other hand, professionals, and the rich, were seen as high in competence but low on warmth, thus, eliciting envy. Drug addicted or homeless people were found to elicit feelings of neither warmth nor competence, viewing them led to feelings of disgust. Furthermore, Amodio (2008) found that besides disgust, insula activity was also found related to intergroup envy. The authors reported a study in which the researchers found insula activity when a disliked out-group member was rewarded.

One further key brain area that was found to be associated with in-group out-group biases is the fusiform face area. The fusiform face area is an area of the brain located within the fusiform gyrus and is strongly associated with the processing of faces (for example compared to objects). Even though earlier studies have suggested that the amygdala might be a key node in intergroup perceptions, later research revealed that the perception of in-group and out-group might be more strongly associated with the fusiform gyrus. This conclusion is based on the results

of studies which have not only investigated intergroup responses of white participants but compared them to responses from black participants. Also Cikara and Van
Bavel (2014) suggested that a limitation of neuroscience research on intergroup
relations might be that mostly Caucasian participants are involved but not African
American participants. For instance, as discussed in Chap. 2, we find that black
participants also show an IAT effect. Furthermore, when conducting fMRI research,
it was found that black participants also showed increased amygdala activity to
black compared to white faces. On the contrary, when considering studies where
fusiform gyrus activity was found, researchers *do* find differences in the processing
of black and white faces, where the activity is the opposite pattern for white and
black participants. This might suggest that the physical identification of race, and
in-group – out-group is strongly related to activity in the fusiform gyrus, whilst the
amygdala reflects emotional processing of social perception. Research has also
found, using EEG that individuals exhibited a greater motor resonance (mu-
suppression) when out-group compared to in-group members made threatening ges-
tures. Furthermore, Van Bavel and Cunningham (2011) predicted that amygdala
activity might reflect responses (threat responses) to novel groups. To test this, the
researchers developed an experiment involving the minimal group paradigm. In the
minimal group paradigm, participants were randomly assigned to groups (say A or
B), and they learned over time who was in in their 'in-group' and who was in the
out-group. In their fMRI study the researchers created mixed race minimal groups,
and found that there was increased amygdala activity to the groups (e.g., viewing a
member of group A versus group B) but that there was no effect of race, thus, the
newly created group membership eradicated the race-effect. In a series of studies,
Molenberghs (2013) investigated the effect of group membership on action perception. First participants were divided into random teams, and then required to press a
button as fast as they could in a competitive game. Later, they had to watch videos
of this button press and judge the performance (how fast they thought someone else
pressed) of in-group and outgroup team members. The authors found that outgroup
button presses were judged as slower, even though the video was manipulated so
in-and-outgroup presses were made equally fast. Moreover, this might indicate
both; (a) that participants judged the performance of outgroup members differently
(but actually truly saw that both were the same), or (b) that participants actually saw
the action differently. Thus, in a second study the authors performed fMRI during
the task and found differential brain activation pattern in the perception but not in
the judgment phase of the task, suggesting that participants literally saw the action
of outgroup members differently.

Another task, which is often used to study prejudice and biases, is entitled the
'weapons identification task'. In the task, a picture of a black or white face is very
briefly shown on the computer screen (i.e., for 200 milliseconds (that is very short)).
Then the target picture – either a handgun or a hand-tool – is presented, and the
participants are asked to categorize the target as a gun or tool, ignoring the face
picture that occurred before. Many studies have found that participants are more
likely to wrongly identify a tool as a gun when they saw a black face before.
Researchers have suggested that participants stereotypically associated black people

with violence and danger. Using this task, researchers have additionally recorded EEG. An EEG measures the sum of electrical activity of cortical areas of the brain with electrodes placed on the scalp. The researchers found that the ERN (error related negativity, which is an EEG component) was different when participants failed to classify correctly (the gun and the tool) after viewing a black face. The ERN might be rated as an indicator of anterior cingulate cortex (ACC) activity, which has been found to be associated with cognitive control and conflict monitoring, thus supporting the view that individuals tend to correct strongly for their biases. Additionally, a further well studied phenomenon in social neuroscience is the own-race effect (Feingold 1914). Commonly, researchers found that people are better at recognising faces from the own race compared to faces from other racial groups. Researchers further found that individuals exhibit greater fusiform gyrus activity to pictures of their own race compared to other races, and that this effect was correlated with the own race effect (i.e., those who showed the highest bias in better remembering faces from their own race, also displayed greatest fusiform gyrus activity to own race faces) (Golby et al. 2001). This might be explained by the fact that people usually have more life experience with in-group compared to outgroup members and also that they might process the faces differently. For example in-group faces might be processed individually whereas outgroup faces may be processed at a subordinate or category level.

In addition, EEG also showed differences in the processing of in-and out-group members. Own race faces also appear to elicit a stronger magnitude of the N170 EEG component (a negative peak in the time-locked EEG signal 170 ms after the stimulus), again this effect can also be observed with randomly newly created minimal groups, and override other cues of categorisation such as race. This might suggest that neuroscience studies support the idea that group formation in general, and competition in particular might fuel prejudice. To summarise; Using fMRI researchers have consistently found increases in amygdala activity when viewing black versus white faces with short presentation duration. However, this effect is not specific to Caucasian participants, suggesting that a general emotional (stereotype) response might be represented here. By contrast, activity in the fusiform face area is highly specific to own versus other race effects, such that faces from one's own race are processed differently to faces from a different race. This effect is present for black as well as for white participants, suggesting that the race of the face (a social aspect) is already processed very early during face processing. This suggest that the categorisation of in-group and out-group also effects the neuronal processing of the faces at an early stage (i.e., around the same time as when a face is determined to be a face rather than an object).

Previous neuroscience studies have also found that interfering with brain activity can have an effect on racial biases. Using tDCS (transcranial direct current stimulation). Sellaro et al. (2015) found that enhancing stimulation over the medial prefrontal cortex reduced implicit racial biases, as measured with the IAT. tDCS is a novel non-invasive simulation technique in which certain areas of the brain can be either activated or reduced. This depends on the polarity, such that anodal tDCS enhances, whilst cathodal tDCS reduces cortical excitability. The initial effect of tDCS can be

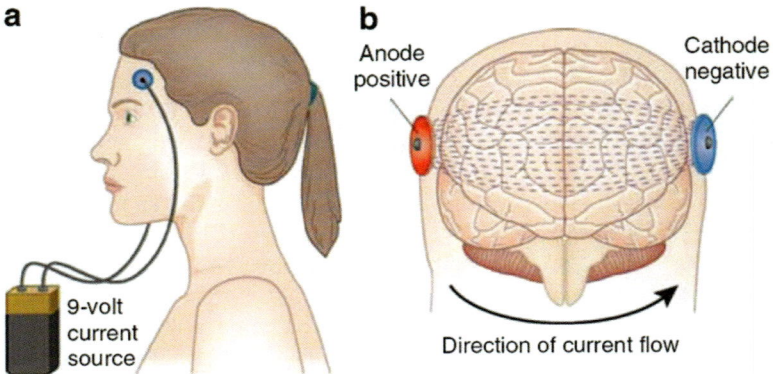

Fig. 3.4 Set-up of tDCS (Reprint permission code: License Number 3910141199835)

explained by effects on the sub-threshold membrane polarisation, and further effects with prolonged stimulation lead to excitatory effects driven by the activity of the glutamatergic system. Figure 3.4 shows tDCS.

In the study, 60 participants received anodal, cathodal, or sham stimulation over the mPFC whilst they were performing an IAT. The researchers found a significant reduction in the IAT with enhancing tDCS. They explained the results by speculating that increasing the excitability of the frontal cortex might have enhanced cognitive control processes over racial biases. This thus supports the role of the mPFC in self-regulation functions. Similarly, other researchers have also found increased prejudice and stereotyping when participants consumed alcohol, which is known to reduce self-control.

However, tDCS has also recently been used for other than research purposes. Figure 3.5 shows an advertisement for at home "treatment" self-tDCS kits, which are claimed to make you smarter, sexier, and happier, by zapping your brain. For example "the brain stimulator" advertises as follow;" What if tDCS could improve your memory, expand your problem solving abilities, or even help you learn new information up to twice as fast?? Researchers have been studying the cognitive enhancement benefits of tDCS for years, and the results they've uncovered are astounding. In other words, tDCS allows you to unlock your brain's true potential!" (See Fig. 3.5.)

Unfortunately, as we have already determined throughout this chapter it is not as simple as that. For example one key problem is location of the stimulation. In a lab experiment one would optimally require a previous brain scan of the participants in order to set the location of the coil precisely. When I watched TMS being applied to a participant, at least 30 min were spent on locating the device, including taking into account various individuals brain scans of the participants, and anatomical mapping software were used in order to decide the correct location. Also effects of tDCS are average effects, and side effects are unclear, thus might not help a single person at all. Furthermore, as discussed before, since we don't have just one area for one

Fig. 3.5 Model posing with public market (not research related) tDCS device (Reprint permission via Permission https://creativecommons.org/licenses/by/4.0)

function, stimulation of one region cannot change a concept. Furthermore effects only last for minutes; only to name a few problems.

Besides neuroscience studies on prejudice and categorisation, recently researchers have also investigated intergroup relations in association with other emotional responses, such as empathy. With regards to empathy, the key neuronal mechanism involved in such processes might be mirror neurons. Those are neurons which are active not only when the participants are performing certain actions themselves, but also when they are observing someone else performing that action. Rizolatti, first determined such neurons within the monkey brain, but now many studies have also investigated the effects of mirror neurons in humans. For instance viewing someone else in pain elicits similar networks (including the dACC) to when the pain is applied to the participants themselves. In a recent study the researchers used TMS to temporally reduce function of the regions of the mirror neuron system, specifically regions of the premotor regions of the frontal cortex, the anterior intraparietal sulcus, and the inferior parietal lobule. Participants then had to either press a series of keys, or imitate a series of key presses from another person. TMS impaired the ability to imitate, suggesting that those brain regions are essentially involved in mirror neuron activity. Indeed, it is a natural human instinct to imitate. We find imitation of course in small babies, and throughout life. Automatic imitation tendencies can also be demonstrated experimentally. For example imagine sitting in the lab and observing a person making certain hand movements, for instance a gesture in which the thumb and the fingers make a U-shape. In one set of trials you are asked to imitate this hand movement, and in another set of trials you are asked to make a different hand movement. It usually takes participants much longer to initiate a hand gesture which is incongruent to the one they are observing suggesting that they have an automatic tendency to imitate. This effect is however reduced for example in patients suffering from autism. However, research also found that the mirror neuron empathy and imitation effect was reduced if the person observed is from an outgroup. And indeed, behavioural as well as neuroscience studies have shown that empathic responses, to physical as well as emotional pain, are reduced for outgroup compared to in-group pictures. For example, it was found that the skin conductance response (a measure of physiological arousal) was reduced for images of black compared to white individuals in pain in Caucasian participants. In addition, fMRI studies found that the brain pain matrix (regions including the anterior cingulate cortex, supplementary motor area, and the insula) were less active when seeing out-group hands being pricked with a needle compared to in-group hands. Furthermore, it has been shown that there are also differences in the extent to which individuals apply theory of mind (thinking about the mental state of others) with regard to an in-group or an out-group member. For example, when thinking about the mental state of another person, activation in the dorsomedial prefrontal cortex can be observed, but is reduced for out-group members. It has also been suggested that the dACC is not only involved in physical pain, but also correlated to experiences of social pain (for example the pain of being socially rejected). For example at UCL, Tania Singer found activity in the dACC regardless of whether the participant was receiving an electric shock, or if they observed their partner receiving an electric shook. In another study by De Wall and Eisenberger, participants took 1000 milligrams of acetaminophen (Tylenol) painkiller medication for 1 week, and a control group took a placebo tablet. During this week participants had to report the amount of social pain that they felt during the day. The researchers found that the painkiller reduced the amount of described social pain. In a further fMRI study they also found brain activation correlates with this effect. During the scan the participants experienced being rejected by another player of a virtual game. This effect of being rejected led to activity in the dACC (as discussed above, the area that is also involved in perception of physical pain) however, the painkiller drug reduced dACC activity. Activity in the dACC (suggestible social pain) was also found when participants observed unfair distributions of points or money between them and another player. Furthermore, it was also found that participants showed subsequent activation in brain areas associated with reward (the ventral striatum) not only when they received money themselves, but also when someone else received money, who had been treated unfairly in the previous distribution of money. This suggests that individuals determine fairness and do not merely act on purely selfish motives. In another study it was found that individuals show activity in the ventral striatum (reward area) not only when they themselves receive money, but even more when they receive only some money but could at the same time give some of their money away to charity. The researchers suggested that humans do feel rewarded (and happy) as a result of giving, and not only by receiving goods themselves. Thus helping others might arise as it is intrinsically rewarding to the person who gives. This might stem from the perception of unfairness (i.e., my friend gets less, my family gets less, people in other countries have less) and a desire to reduce such inequality if it is perceived as unfair, which can then lead to rewarding experiences. This does not imply that everyone is happy giving their money to charity, but it might explain why people help others and are social rather than being purely selfish. Indeed, some people might help their family, some might help a charity to support poorer countries, others might help a friend at work, some might help a foundation to provide help for research or the environment. In any case, compassion for the cause might induce reward after the help.

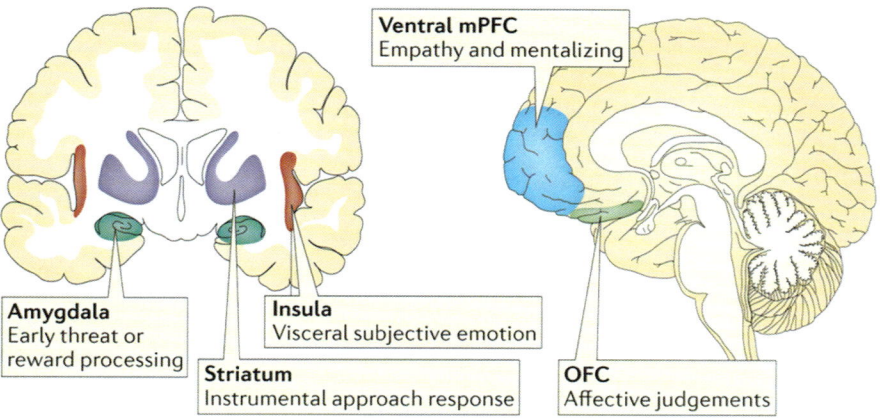

Nature Reviews | Neuroscience

Fig. 3.6 Taken from Amodio (2014). Prejudice Network. An interactive set of neural structures that underlie components of a prejudiced response. The amygdala is involved in the rapid processing of social category cues, including racial groups, in terms of potential threat or reward. Approach-related instrumental responses are mediated by the striatum. The insula supports visceral and subjective emotional responses towards social ingroups or outgroups. Affect-driven judgements of social outgroup members rely on the orbital frontal cortex (OFC) and may be characterized by reduced activity in the ventral medial prefrontal cortex (mPFC), a region involved in empathy and metalizing (Reprint permission reference License Number 3898260809604)

Figure 3.6 shows the "prejudice network". In his review article, Amodio (2014) established key brain areas that were found to interact in processing during intergroup stimuli. Before in this chapter, I have already discussed the function of each of these brain areas. Amodio further describes the function of the amygdala, being involved in the processing of facial expressions of fear, and thus integral to the rapid processing the potential social threat. Researchers also found in a more recent study, involving black participants that the amygdala might habituate more slowly to outgroup faces. However, as noted before previous studies also found that it was not race as a factor per se to elicit amygdala responses, but rather group status. For instance studies found that if t-shirt colour predicted group membership than t-shirt colour and no longer race predicted amygdala activity, such that the activity for outgroup colour t-shirt elicited amygdala activity regardless of race. However, Amodio also suggested that the amygdala activity might not reflect a threat response to the face but rather the fear of appearing prejudiced.

Derks et al. (2013) edited a book, which includes many research articles on the neuroscience of intergroup relations, involving a variety of methods such as fMRI, EEG, and physiological measures. After examining the book it becomes clear, that prejudice involves many complex and interrelated processes and thus a network of related brain area activation patterns. In 2013 Molenberghs published a review article on the neuroscience of intergroup relations. He described different stages, or components, involved in processing of intergroup stimuli. In particular he names stages of social categorisation, action perception, empathy, and face perception. For instance with regards to social categorisation the authors mentions a study where it was found that the medial prefrontal cortex might also be involved in group membership processing. Specifically, in the study, participants were divided into random groups (a blue or a yellow team); subsequently they divided money to themselves, their in-or-outgroup members. During trials in which participants showed in-group favouritism (i.e., they gave their own group more) medial prefrontal cortex activation was found. Posterior anterior cingulate cortex activity however, was found in trials of social comparison, suggesting that different brain activity patterns might mediate social identity and social comparison.

Taken together, neuroscience studies seem to support theories developed in behavioural experimental social psychology, as discussed in Chap. 2. Specifically, neuroscience studies have supported the approach that empathy might be reduced for individual's outsides one's own group. However, until recently, neuroscience studies investigating intergroup relations have not included studies that utilised psychopharmacological techniques, which our own research was one of the first to employ. Before discussing this research, the basics of psychopharmacology will be described.

3.2 Basics of Psychopharmacology

Psychopharmacology (Psyche = mental state; Pharmakon = drug) involves researching drugs that influence human mental states. Pharmacotherapy is indeed the most commonly used treatment in psychiatry. Psychoactive substances and their synthetic derivatives are used to treat a variety of mental disorders. The development of pharmacological treatment for mental disorders was a great advance. Historically, people did not know how to respond or help people with mental health problems, which therefore led to the inhumane treatment of such people. For example, in the nineteenth century patients were locked in houses for so called "mad men". The pubic would visit those "mad men" laughing at them as a Sunday afternoon family entertainment. Treatments similar to torture were used, which was partly due to the fact the people did not know how else to respond. Drug treatment which helped some patients to recover and lead a normal life thus improved the situation. Besides treatment with drugs however, we now understand that a combination of pharmacological with psychological therapies and education can be most beneficial, but this also depends on the individual patient (for example, which particular drug, and dose of the drug, or combination of drugs, and which psychological treatment might be best). However, it should be noted that drugs do not help all patients. For instance in schizophrenia, only about 1/3 of patients fully recover after psychotic episodes when given treatment with drugs. Today we often hear that drugs might sometimes be prescribed too easily. For example we read in the news that children who might not meet the full criteria for ADHD are prescribed ADHD medication nevertheless.

Or that anxiolytic drugs as well as anti-depressants are prescribed even though a strictly diagnosed mental illness such as depression or anxiety disorder is not present. Whilst some psychiatrists might argue that drugs should only be prescribed if a diagnosis is strongly supported, and if impairment in social or occupational settings is present, others might argue that it might be difficult to establish a diagnosis with a 100 % certainty. For instance unlike some physical problems (I definitely have broken my arm), there is no such definite test for mental disorders. It is unquestionable that individuals that seek help need help, but it might need to be discussed if this should be in the form of drugs. In Chap. 5, I will discuss that some might argue that drugs are not different to other forms of intervention (i.e., both affect the brain). First of all, however, I will explain the biological mechanisms of drug action.

Many drugs affecting the brain interact with the action of neurotransmitters (the chemicals in the synaptic cleft between two neurons, responsible for neuronal signal transport). Neurotransmitters are endogenous chemicals with each neuron releasing one 'main' neurotransmitter; for example, the neuron is cholinergic, dopaminergic or glutamatergic. Some of the most important neurotransmitters are acetylcholine, serotonin, dopamine, noradrenaline, adrenaline, glutamate, and GABA. For example, acetylcholine is involved in regulating muscle movement, and can be found in motor neurons. However, again there are multiple other functions of acetylcholine, such as regulating pain, sleep, memory and learning. Damage to the cholinergic system is associated with Alzheimer's disease. Dopamine is also involved in multiple functions and a deficit in dopamine is associated with Parkinson's disease (which involves motor dysfunction). Serotonin is involved in regulation of mood, and glutamate is the main excitatory neurotransmitter in the brain. That is a very simplistic picture here; reminding you that I wrote a review article about ONE – of many – glutamate receptor (mGluR5) and its involvement in anxiety, depression, addiction, learning and memory etc. etc. Thus, even though one neurotransmitter might be predominantly involved in a certain human function, the complete action of one neurotransmitter is vast, being involved in many different processes.

The mechanism of action of drugs depends on multiple factors. Oral administration of drugs – unlike intravenous – may require metabolic activity for the active form of the drug to reach the blood stream. Drug absorption depends on age, gender, body size, and other factors. Also only lipophilic (fat loving) drugs can cross the blood brain barrier and produce an effect on the brain, whilst others act only peripherally. Drugs can bind to enzymes, membrane carriers, ion channels, and receptors. Drug receptor complexes have certain characteristics. One is specificity, which means that the structure of the ligand must conform to the 3D structure of the receptor (for example metabotropic glutamate receptor number five has different structures than glutamate receptor number two). Thus, when discussing a drug that affects noradrenaline then strictly speaking this means that it interacts with *some* noradrenaline receptors. Another factor is receptor population. The drug effect also depends on the number of 'available' receptors to interact with.

Psychotropic drugs (i.e., those acting on the central nervous system) can be broadly classified into antipsychotics, antidepressants, anxiolytics, mood stabilizers, prescriptive stimulants, and sedative-hypnotics. Examples of antidepressants the (most-commonly prescribed) are Selective Serotonin Reuptake Inhibitors (SSRIs), such as fluoxetine. The exact mechanism of action is still unclear, but it is understood that the drugs inhibit the reuptake of serotonin in the synaptic cleft, thus allowing more serotonin to bind with 5-HT receptors. Anxiolytics and sedatives such as benzodiazepines enhance the activity of the inhibitory neurotransmitter GABA. Lithium is a mood stabilizer, often used for severe manic-depression, which interacts with numerous processes linked to neurotransmission, has strong side effects and a high toxicity profile. Stimulants include for example, various amphetamines which are powerful releasers of dopamine and noradrenaline. Drugs can produce so called adverse drug events, which are unexpected, undesired, or excessive responses to a drug and lead to the need to change the drug, or modify the dose. However, this can also result in temporary or permanent harm, disability or even death of the patient. General adverse drug effects include: allergic reactions, drug-drug interactions (many drugs interact adversely with alcohol), cardiac disturbance (e.g., alarming drop in blood pressure), fainting or dizziness, jaundice (liver dysfunction) or unusual bruising. It is estimated that side-effects of drug treatment alone costs the National Health Service around £ 5.6 million per year.

The drug, propranolol, which I used in our studies on intergroup relations, affects receptors for the noradrenaline neurotransmitter. Noradrenaline is synthetized from the amino acid tyrosine. Noradrenaline is also the primary neurotransmitter in peripheral sympathetic nerves and involved in heat rate regulation, blood circulation, and respiration rate. Propranolol blocks noradrenergic beta 1 and beta 2 receptors in the body and in the brain, thus reducing activity of noradrenaline. Also propranolol has many side effects listed, some of the common (more than 1 in 100 people) others rare (more than 1 in 10,000 people) including, sleeping problems, tiredness, gastrointestinal problems, blood problems, fainting or brief loss of consciousness, skin rash or rashes, worsening of circulation problems. In Chap. 4, I will describe how and why propranolol might change how we see other people.

Open Questions Chapter 3
• Can neuroscience provide a helpful tool to study intergroup relations?
• What exactly was found in the Phelps study?
• What is the future of neuroscience?
• Can people *read your brain* using fMRI?

References

Amodio, D. M. (2008). The social neuroscience of intergroup relations. *European Review of Social Psychology, 19*, 1–54.
Amodio, D. M. (2014). The neuroscience of prejudice and stereotyping. *Nature Reviews Neuroscience, 15*, 670–682.
Cikara, M., & Van Bavel, J. J. (2014). The neuroscience of intergroup relations: An integrated review. *Perspectives on Psychological Science, 9*, 245–274.

Cunningham, W. A., Johnson, M. K., Raye, C. L., Gatenby, J. C., Gore, J. C., & Banaji, M. R. (2004). Separable neural components in the processing of black and white faces. *Psychological Science, 15*(12), 806–813.

Derks, B., Scheepers, D., & Ellemers, N. (2013). *Neuroscience of prejudice and intergroup relations*. Psychology Press.

Feingold, C. A. (1914). The influence of environment on identification of person and things. *Journal of Criminal Law and Police Science, 5*, 39–51.

Golby, A. J., Gabrieli, J. D. E., Chiao, J. Y., & Eberhardt, J. L. (2001). Differential fusiform responses to same-race and other-race faces. *Nature Neuroscience, 4*, 845–850.

Lieberman, M. D. (2003). A social cognitive neuroscience approach. In *Social judgments: Implicit and explicit processes* (5th ed., p. 44). Cambridge: Cambridge University Press.

Lieberman, M. D. (2013). *Social; Why our brains are wired to connect*. Oxford: Oxford University Press.

Molenberghs, P. (2013). The neuroscience of in-group bias. *Neuroscience & Biobehavioral Reviews, 37*, 1530–1536.

Phelps, E. A., O'Connor, K. J., Cunningham, W. A., Funayama, E. S., Gatenby, J. C., Gore, J. C., et al. (2000). Performance on indirect measures of race evaluation predicts amygdala activation. *Journal of Cognitive Neuroscience, 12*(5), 729–738.

Restak, R. (2004). *The new brain*. London: Rodale.

Sellaro, R., Derks, B., Nitsche, M. A., Hommel, B., van den Wildenberg, W. P., van Dam, K., et al. (2015). Reducing prejudice through brain stimulation. *Brain stimulation, 8*, 891–897.

Van Bavel, J. J., & Cunnigham, W. A. (2011). A social neuroscience approach to self and social categorisation: A new look at an old issue. *European Review of Social Psychology, 21*, 237–384.

Chapter 4
Psychopharmacology and Prejudice

In the previous chapters I have described the social psychology and neuroscience in general, in this chapter I would like to introduce studies in more detail in order to describe the complexity as well as prospects of such work. I will to introduce our studies which have been widely discussed in the media as apparently suggesting a cure for racism. The 2012 paper, "Propranolol reduces implicit negative racial biases" received the Springer Neurostar award for being the most discussed paper in the media/social web of all neuroscience Springer journal articles in 2012 worldwide. In Chap. 1, I introduced this book, discussing examples of prejudice, starting with the extremes of prejudice as exhibited by the majority of Nazis during the Second World War. Many examples from the literature about racism, or any form of prejudice, and its negative consequences on the victims and on society have been given. Chapter 2 discussed the social psychology of prejudice. What is prejudice? What is racism? This is a highly important chapter, as a definition of a concept is essential in understanding psychological investigations. If a headline reads "cure for racism found", this would seem to be obvious terminology, but at closer inspection, one recognises the importance of understanding how psychologists and others measure prejudice and how concepts are defined.

Equally important was Chap. 3, in which neuroscience and psychopharmacology are discussed. It will have become clear by now, that the studies that will be discussed in this chapter, involve "drugs and prejudice" together. Therefore, one has to understand how these drugs work, what the basic principles of neuroscience are, and indeed the limitations of this field. However, sometimes neuroscience popular science books and more general articles about neuroscience lack this information. Many books on neuroscience start introducing the topic by stating how research has developed and that modern technology now allows us to view the brain in action. That we can 'almost' detect what is happening in the brain if someone is lying, if someone is happy, or if someone is sad. Indeed technological development and advancement along with fMRI research, has enabled progress in the field of understanding, but there are also limitations. After reading Chap. 3, one might have realised that neuroscience research is vastly complex. For example, the results

© Springer International Publishing Switzerland 2016
S. Terbeck, *The Social Neuroscience of Intergroup Relations: Prejudice, can we cure it?*, DOI 10.1007/978-3-319-46338-4_4

depend on how the concept was measured, which stimuli were used, which scanning parameters, etc. etc., etc. Similar issues apply to the field of psychopharmacology. How simple and appealing sounds the idea that one is always happy when serotonin levels are high and that one can take certain drugs and puff, like magic; whatever problem one has is gone. However, it is more complicated than this. Chapter 3 also illustrated the complexity of biochemistry of drugs.

Therefore, it is important to understand some of the complexity of the fields as progress might only be made if the current state is fully explained and understood. Indeed, it would have been much easier to discuss the negative effects of prejudice, and that there was now this magic drug that could cure it. However, it isn't that simple. Realistically, if there was a magic drug for pure happiness or to prevent aging or to produce peace and love would we not all use it by now? If there was a drug that could just make you fly and nothing else, with no side effects, would one not take it? This chapter will first discuss the study, "Propranolol reduces negative implicit racial biases" and then proceed to discuss the study "Beta-adrenoceptor blockade effects on fusiform gyrus activity to black versus white faces." The intention of this program of research was to determine whether emotional arousal was involved in prejudice. We were looking for a robust way to block emotional arousal, without also affecting many other psychological variables. As already discussed in Chap. 2, with regard to prejudice, behaviour as well as neuroscience research in social psychology, has suggested that some forms of racial biases may have a strong emotional component. For example, it was found that the IAT effect (the measure of implicit racial biases) was correlated with activity in the amygdala, a brain region that is highly associated with emotion processing. Furthermore, behavioural interventions, such as increasing intergroup contact between groups led to participants reporting that they felt less negative about the other group. However, as discussed previously, we do not know which emotion may be involved in implicit biases. For instance, an obvious candidate might be fear, an idea which is also reflected in everyday language. For example, in the English language prejudice is often also described in implying fear, such as "Xenophobia", or "Homophobia". The term phobia suggests that fear might be a central emotion in racial biases.

Other than fear, aggression might also play a significant role. For example, in other non-English speaking languages "Xenophobia" translates to words that are associated with hate or aggression, rather than fear, and in the Chinese is related to disgust. In German, the word for Xenophobia is "Fremdenfeindlickeit", which translates to; "hate of foreigners". Besides what we already find in everyday language, numerous studies in social psychology as well as neuroscience (See Chaps. 2 and 3) support the idea that emotions might be involved in prejudice; more specifically, that emotions might additionally be a stronger component of implicit compared to explicit prejudice. Thus, the idea of the first study was to provide further evidence for this theory. The idea for the study was, as one might say, purely scientific; to understand the psychological mechanisms that operate when people are prejudiced. Indeed, I was not looking for a drug that might cure prejudice; a la 'Some researchers are looking for a drug to cure cancer, I was looking for a drug to cure prejudice'. Of course most of us agree that it would be great to reduce prejudice, but the idea behind the study was to investigate the causal role of emotion in implicit racial biases. It is important to note that I have just written the "causal role" because many previous studies described were correlational and did not suggest specific causal relationships.

Consider for example, a survey in which participants are asked two questions; (a) How many times contact they had with people from a different group, and (b) How much fear and aggression they experience when thinking about people from that group. If someone had high intergroup contact and they also had low intergroup anxiety, then this might mean that intergroup contact reduces intergroup anxiety but it could also be the other way round; that because people were low on intergroup anxiety they searched for more intergroup contact. fMRI is also a correlational method, for example, we say that "activity in the amygdala correlates with the IAT", we cannot say that the amygdala activity caused the IAT effect. Indeed, this would be a circular claim. Consider for example that one says that they felt anxious because their amygdala was active, it could just be that their amygdala was active because they felt anxious. In order to make causal claims, one needs an experimental design, such as a psychopharmacological experiment along with a behavioural experiment in which variables are set and manipulated, deliver. Coming back to the relationship between intergroup contact and anxiety; if a researcher were to put half of the subjects in a group setting in which they meet people from the other group and the other half of people in a group setting where they did not meet people from another group, and the researchers measured intergroup anxiety before and after this intervention, then one could make a causal claim. For example, if intergroup anxiety was lower in the group that had mixed intergroup contact, then we could say that contact *caused* the anxiety to be reduced. Thus, if one gives a group of people a drug and another group a placebo, and then measures changes in behaviour, everything else being equal, one could say that the drug *caused* the changes.

As stated before we were interested in finding a method to reduce emotional arousal, in order to see if this would reduce implicit racial biases. In psychology, there are many ways to experimentally manipulate emotions. These procedures are called mood induction techniques. Using a drug might be described as one method of mood induction. There are other methods of mood induction; for example, one well established method involves participants watching or reading an emotional story or film. This is an obvious method, you watch a horror film, and you are anxious. Shall I say it? Yes, it is more complicated than this here too. This now seems to be a most obvious sentence. If I wanted to induce mood, say increase fear, why not let participants watch a horror movie? Participants watch a horror movie; they are anxious; obviously right? I will briefly discuss why this is more complex and what advantage using a drug as a mood induction method might bring. Imagine that you were to conduct an experiment in which you wanted to increase participants' fear by having them watch a horror movie. Firstly, you would need to consider; which movie? Would you take a movie that people might have seen before? Would they maybe be less anxious if they knew what would happen? Secondly, is everyone getting anxious? Or might some even find it funny? Are they only getting anxious, or are other emotions involved? Some people might also get angry, or depressed?

Imagine you now found a new movie which apparently made people very anxious. How will you measure this?

One could ask about their anxiety (but should also ask about their aggression, happiness, disgust, sadness level) before and after they watch the film. Imagine it worked. 20 people watched the film and most of them say they felt anxious afterwards. There are however, at least two problems with this; (a) experimenter demand effects, and (b) time course of the effect. Actually, there is a third problem, which is (c) specificity of the emotion (but that is also a problem for drug studies). Experimenter demand effects are effects that are observed when one suspects that the participants detect the hypothesis of the experiment and answer questions in accordance with it, in order to please the experimenter or to comply with what they think might be expected of them as participants. For example, if participants were asked to fill out a questionnaire, stating how anxious, aggressive, sad, happy, etc. they felt and they were then shown a horror movie followed by the completion of the mood questionnaire again, it becomes obvious to participants what would be expected. Participants would mostly know that watching a horror movie makes you anxious and they might – consciously or unconsciously – report that it made them feel more anxious even if it did not. One way to overcome this problem is to use other methods than self-report questionnaires. For example, when someone is anxious they will also have an increased heart rate.

Thus, one could investigate the effect of the movie on the levels of anxiety using physiological measures to investigate anxiety, which is exactly what previous researchers have done. In a recent meta-analysis of traditional mood induction techniques (in which they combined the results of a large number of experiments that used movies to induce mood), researchers found that changes in emotion are mostly detected on self-report questionnaires and not on physiological measures such as heart rate. This would indeed suggest that the mood induction using a film might be less efficient as it produces changes that might be related to experimenter demand characteristics. Or the mood induction using such methods might simply be very mild – changing the mood only very little – so that it did not result in physiological changes. Problem (b) is the time course of the effect. Not many researchers have investigated this issue, but it could be suspected that the mood changes of watching a horror movie might be very temporary. Indeed, participants could be anxious during the film, but then this anxiety would be completely gone when the film is over. However, the mood change might also last longer than this, but how much longer? And is the duration the same for everybody? The third problem about the specificity of mood is a problem that any known method – including drugs – has.

To be precise, it might not be possible to only investigate fear or sadness. Perhaps this is also due to the idea that human emotions are not as simple as that to start with. In many situations one might feel predominantly fear, but one might at the same time also experience sadness and aggression. In this regard, traditional mood induction methods, such as watching a film might have an advantage, because indeed, watching a horror movie might make people predominantly anxious, whilst drugs often effect emotional arousal in general, and are much less specific than other methods. However, with regard problems (a) and (b) using a drug may have its advantages (though it has other problems). For example when considering the experimenter demand effect, certain drugs produce strong significant changes, which outweigh only perceived changes. The drug we used reduced heart rate, blood pressure, and changed emotions in the majority of participants. This can't be taken as a general claim, but one might say that drugs might sometimes produce stronger changes in mood than watching a film or reading an emotional story. Secondly, the time course of the effect is clearer; drug effects, or mean drug effects, can be time locked. So for example, previous research indicates when the effect of the drug is strongest (for example after 30 min) and when it reduces again (for example after 2 h). This allows planning the experimental study in a way that the psychological tests are conducted at the time of maximum effectiveness of the drug.

Before I proceed to further discuss the specific drug that we used in this study (propranolol) I will briefly review the methodological problems with using a drug as a mood induction technique, and also why propranolol was the drug of choice. There are two further problems with pharmacological mood induction; (a) the effect strength and selectivity of the effect and (b) the complexity of the physiology. Consider problem (a). I will never forget how a friend of mine once told me "So participants took a drug, were tired, half dead, crawling on the floor, but not prejudiced? Great." With many drugs the effect can indeed be quite strong and also produce numerous side effects. Say if I was to take a large dose of Valium, it might have a strong effect on me, and also maybe make me unable to complete an IAT at all. Propranolol has the advantage that it produces subtle changes in emotional arousal without impairing other physical or mental functioning too much. However, it has to be stated, that even propranolol, or indeed any drug, will always have an effect on more than one specific mechanism. As already discussed in Chap. 3, the activity of neurotransmitters are non-specific, and even though drugs might have predominant effects on certain psychological mechanisms more than on others, there is no drug that simply effects *one* isolated variable. This already leads to problem number two, which is that the drug action might be unclear. For example, propranolol predominantly affects the neurotransmitter noradrenaline, but noradrenaline will again interact with dopamine, serotonin and hormones. Thus, even though the effect of the drug might be mostly restricted to the activity of one neurotransmitter, the whole picture of the effect is more complex than this.

In our study, mood induction – using the drug propranolol – was employed to reduce emotional arousal in order to investigate its effect on implicit racial biases. As described before, propranolol is a beta-blocker which affects the central and peripheral activity of the neurotransmitter, noradrenaline. When discussing our research, I often meet people telling me that they are taking beta blockers themselves. Indeed, beta blockers are – or at least were until very recently – widely prescribed drugs for a variety of different medical and psychological conditions. The first line treatment that beta blockers are prescribed for is hypertension (high blood pressure), although currently medical doctors in the UK prefer other treatments such as various kinds of angiotensin inhibitors for hypertension rather than beta blockers. In addition to the treatment of hypertension, there are other conditions for which beta blockers are prescribed. One is acute anxiety, specifically performance anxiety. For instance, violinists, who suffer from high performance anxiety during music concerts, would not be able to play their instrument if their hands were shaking. Beta blockers can lower anxiety by reducing physiological and psychological symptoms of stress responses. When being anxious or highly emotional aroused the sympathetic nervous system is active. We experience for example, increased heart rate, as well as increased sweating (i.e., wet hands). These reactions are part of the activity of the sympathetic system. The sympathetic activity is mediated by the neurotransmitter noradrenaline. Very early studies showed that after perception of danger, an immediate automatic stress response is elicited; the fight or flight response, leading to the described symptoms of acute fear. As discussed, this response is partly mediated by the activity of noradrenaline, which acts on beta receptors, such as in the heart. The beta blocker drug interferes with the beta-receptors (See also Chap. 3.) so that the physiological and behavioural responses to the fearful stimulus cannot be elicited. Therefore, the physiological symptoms of the fight or flight response are reduced. Besides physiological symptoms, anxiety also produces a unique feeling; such as being nervous, feeling of wanting to escape, etc. The 'feeling part' might be mediated by central activation and partly by central noradrenaline transmission in the brain. However, not all beta blockers act centrally, some cannot cross the blood brain barrier and thus only act on peripheral symptoms. However, propranolol is able to act centrally as well as peripherally. Recent research, using fMRI has shown that the amygdala might be a key node involved in noradrenergic processing. In the study, the participants either received propranolol or a placebo before viewing emotional pictures (such as a scary monster, a neutral chair, a woman screaming, someone crying) during the fMRI scan. Propranolol was found to reduce the activity of the amygdala to such emotional pictures. It was however not clear, that anxiety was the predominant (or at any rate the only) emotion affected. More likely is that propranolol reduced peak arousal of any emotion. Indeed, there are some studies suggesting that propranolol might also reduce aggression.

A second condition for which beta-blockers are currently investigated is Post-Traumatic Stress Disorder (PTSD). PTSD is a mental disorder in which individuals are troubled by recurring, anxiety provoking thoughts and episodes which contain memories of extreme stressful past experiences. For example soldiers, who have been faced with catastrophic and life threatening situations during war, could experience flash-backs as well as generally heightened arousal after such traumatic experiences. The idea is that beta blockers, when taken before or immediately after the traumatic events reduce the emotional arousal experienced, and might thus prevent the development of PTSD. Indeed, recent research has shown that beta blockers can reduce the effects of traumatic memories and memory consolidation by reducing the arousal associated with such events. In conclusion, noradrenaline is involved in fight and flight responses, which are associated with heighted emotional arousal. Beta blockers, such as propranolol, block the activity of noradrenaline, by blocking certain types of beta receptors, thus, reducing noradrenergic transmission. Therefore, propranolol reduces emotional arousal, such as anxiety and other emotions. If emotional arousal is significant for certain forms of racial biases then propranolol should reduce them.

Our propranolol and racial bias study was conducted in the summer and autumn of 2011 at Oxford University. Drug studies require different set ups than regular experimental studies, whereby one needs extended ethical approval; participants need to be more fully informed (not just via reading an information sheet) and there also needs to be a full medical and psychological screening before the experiment, as the participant's safety is paramount. Participants sign up to a screening visit after they see the advert of the study at the university homepage, or in a local newspaper (thus not all participants will be students). Then via email they received a detailed information sheet, explaining that they will be given propranolol or a placebo and that they will be completing a number of psychological tests afterwards. Not everybody though reads the information sheet. I had participants turning up for the screening visit saying: "What, I'll be taking a drug?" Thus, an oral presentation, informing them about the study is usually given in addition during the screening visit. The screening visit takes place on a separate occasion to the actual experiment and is usually conducted by the researchers and a medical doctor, who reviews the medical suitability of the participant for the study. No drug is free from side effects and so therefore, participants need to be informed about all possible side effects of the drug that they might be taking. Propranolol is said to be a relatively safe drug, but as stated above, can lead to tiredness, sleep problems and other side effects. It is particularly important to ensure that the participants have no contraindication to taking beta blockers, as this could produce harmful effects. For example, people who have low blood pressure, or epilepsy, or a previous allergic reaction to propranolol would be excluded from the study. We also conducted an ECG (a measure of the heart's rhythm and activity) to only include participants, where no abnormal pattern on the ECG was detected, and whose blood pressure was within the normal range. Further, as propranolol might interact with other substances, participants were excluded if they were taking any other medications, if they were dependent on alcohol or drugs or if they smoked more than a few cigarettes a day. As we wanted to recruit medically and psychologically healthy volunteers, participants with a history of significant medical or psychiatric problems were also excluded. These participants were excluded before the experiment; i.e., at the screening visit we did not judge it appropriate for their own medically safe to take part in the drug experiment.

The actual experiment was conducted in the morning. Participants were instructed to eat only a light breakfast and not to drink alcohol the night before. When they arrived they were seated in the experimental room and their heart rate was measured. The normal pulse rate varies between 60 and 100; "The lower the rate the fitter is the individual." Indeed, high power athletes could have resting heart rates of about 40–50. A young, healthy female, would usually have a resting pulse rate of about 70. Measuring heart rate, as a discussed before, is the physiological manipulation check. Heart rate was measured throughout the study every 30 minutes in order to note if indeed propranolol reduced physiological components of arousal. More specifically, heart rate was measured. Besides heart rate, measures of self-reported emotional arousal were administered. This was performed four times during the experiment, as well as the first time before they took the drug or the placebo.

Participants rated on a scale, how anxious, happy, relaxed, aggressive, sad, etc., they felt at that moment. The study was a double blind study; meaning that neither the participants nor the experimenter knew if the participant had taken propranolol or the placebo. Only the medical doctor knew.

The placebo (a sugar tablet) or propranolol were put into identical capsules, and then placed in sealed envelopes, with a separate code, so that only the medical doctor, who had the code list knew which envelopes contained the propranolol and which contained the placebo. After participants had completed the baseline measure of heart rate and had completed the self-report mood questionnaire, they took the drug or placebo, and then waited. Admittedly, I personally would find it quite hard to just wait. I would probably constantly monitor and ask myself how I felt, and if the drug effect had already taken place. I might even imagine that I felt relaxed, and that my heart rate was low, all those effects that participants were told that propranolol has. I guess I might be a good candidate for placebo effects. That is indeed why studies involving drugs are placebo controlled studies and not studies in which participants receive the drug or nothing (and placebo effects can be very strong). Often people may define placebo effects as the effects that one imagines even if they don't get a real treatment, but that is strictly speaking not quite correct. Indeed, a placebo is an inactive substance (a sugar pill) and it is often observed that taking such a pill can produce effects. But what is questionable about the statement just made before is that participants imagine those effects (i.e., that they are not real). This is because research on placebo-effects showed that placebos can indeed produce very real (also physical observable) changes in the person.

The most well researched example of placebo effects are studies looking at placebo effects on pain perception. If one experiences pain and then receives a placebo drug or a placebo cream that apparently includes an active substance to reduce the pain, then usually some participants report that they feel less pain. However, beyond this self-report, recent fMRI studies have shown that the activity in pain related brain areas is also reduced. In another fascinating placebo effect study, the participants (all cleaners) were divided into two groups. Group one was told nothing, and group two was told how good their cleaning was for physiological health with cleaning being equivalent to doing exercise. Apparently the two groups then changed nothing in their cleaning routine but after two weeks, the group that was told that cleaning was like exercise, lost weight. We recently also conducted a study, which involved placebo effects. Part of this study was to investigate participants' ability to concentrate. In this study participants received a placebo pill but were that it was a special vitamin supplement which had the effect of increasing concentration. After this, participants completed a test in which they had to circle mixed up numbers in the correct order (for example, bringing '3 – 5 – 1 – 2 – 4' in the correct order as fast as possible) a well-established test of concentration. Those who received the placebo pill that apparently increased concentration indeed concentrated better, and were faster at this task. These results also confirmed a previous study in which healthy seniors received a placebo pill that was described as improving their cognitive abilities. In this study, seniors with the placebo showed enhanced concentration and other cognitive abilities. Another area where placebo effects might play a role is food. Yes, food. Consider "Why turkey and bananas can make you happy." As discussed in Chap. 3, serotonin has been associated with mood regulation, and it was suggested that low levels of serotonin contribute to negative mood such as anxiety and depression. There is one method to investigate serotonin function without actually giving a drug, but by using the amino-acid depletion paradigm. The amino-acid, Tryptophan, is the precursor of serotonin and the amount of serotonin in the brain depends on the amount of tryptophan that is available to it. Tryptophan is consumed within normal food. In fact, humans eat tryptophan all the time. In addition, one can also buy Tryptophan amino acid in health food stores as a food supplement. There are particularly high levels of Tryptophan in turkey and bananas, making me, and probably others – wrongly – believe that eating lots of turkey would make you happy. Why this does not work is because the level of tryptophan compared to others amino-acids that are absorbed with everyday food (i.e., bananas and turkey) are not high enough to produce any noticeable difference. So if eating turkey makes one truly happy (e.g., at Christmas) then this might be a placebo effect. However, in research studies, serotonin function is indeed investigated using amino-acid paradigms, but the mechanisms here are quite different. Rather than consuming the particular amino-acid, the study involves an amino-acid *depletion* procedure, in which one takes all amino-acids but not Tryptophan, which subsequently reduces net brain concentration of tryptophan and thereby, serotonin. More specifically, after fasting for 24 h, participants drink a shake that contains all amino acids but not tryptophan, and then complete the psychological tests. The control group does the same – they eat nothing for 24 h – and then drink an amino-acid drink that contains all amino acids, including Tryptophan. Research has shown that participants in the experimental group are more depressed, aggressive, and anxious. Also, note the fact that both groups were hungry and drank a weird shake after eating nothing for 24 h, so it can't be down to this. This thus supports the role of serotonin in emotion regulation, or emotion elicitation, and in particular negative emotional arousal. To sum up, psychopharmacological studies are placebo controlled because one needs to control for the effect of placebo, so that the observed effects on psychological tasks are not due to the fact that participants are thinking about the changes that drugs might produce, or to the fact that they are taking a drug, but that they are due to the active substances in the drug.

Reading about placebo effects might have taken you 30 min. Thirty minutes was also the time the participants in the propranolol study were waiting and reading or working, before I returned, to again check their heart rate, and to give them the second mood questionnaire. Then they were left to wait for another 30 min before the experiment started. Propranolol has a peak blood level concentration and activity 1h after drug intake, but initial effects might be noted after 30 min. However, as noted before, the effects of propranolol are quite subtle, and even though heart rate would be reduced on average, many participants, taking the placebo pill experienced a slight reduction in heart rate, possibly due to the mere fact that they were relaxing and reading quietly in a room. Besides this, a dose of propranolol of 40 mg is relatively low. Propranolol has previously been used in experiments with doses of 80 mg. A dose of 40 mg is about equivalent to what GPs would normally prescribe for the treatment of hypertension, to be taken three times a day. As every participant received the same dose we also needed to control for body weight, as greater or lower weight might interfere with the effect. Thus, in addition, during the screening we only selected participants whose body mass index was within the normal range, so they were not over-or underweight. Since we gave everyone the same dose we could not measure dose dependent effects, which are optimal in psychopharmacological research. Dose dependent effects mean that the drug effect correlates with the drug dose, so that the higher the dose the stronger the effect. For example, an anxiolytic drug might make people less anxious, and moreover the higher the dose the less anxious they feel. After one hour waiting, completing the mood questionnaire and the heart rate measure again, participants were brought to the experimental room, where they would complete an implicit bias test.

In Chap. 2, I discussed the Implicit Association Test (IAT), which the subjects completed first. The test takes about 20 min. When they finished they had another heart rate measure and then completed the explicit prejudice questionnaire. In particular, they first completed the feeling thermometer. The feeling thermometer is a paper and pencil ranking test of different social groups. First, participants indicate how warm they feel towards their own group (in this study all subjects were Caucasian). They rate this – analogue to a thermometer – from 0 degrees (= very cold) to 100 degrees (= very warm). After this, they rated how warm they feel towards other groups, such as black people or Asian people. After they completed the questionnaire, they would have another measure of heart rate. At the end of the experiments participants were told more about the hypothesis of the study. Before this though, the medical doctor would tell them (but not me) if they had taken the drug or the placebo. Some people discussed general issues of prejudice with me. For example, I often heard participants asking if they were now "racist" as they might have noticed that they found the IAT quite challenging, and they might have found it much harder to sort good words to black than to white. I told them that they were not racist, and basically told them all about implicit and explicit attitudes and the relation to behaviour (See Chap. 2). The effect of propranolol lasts for about 2 h, and participants were advised not to drive but were paid for using public transport, as well as a fee for their time and effort. Participants reported that they enjoyed taking part in the study.

The IAT analysis is completed following an algorithm which the inventor of the IAT – Prof Tony Greenwald –created. There are certain steps one follows, for example, excluding trials that have impossibly long or short response times, or that contain a high number of errors. All the rules for this (i.e., which response times are rated as too high etc.) are fixed by the algorithm. After this initial step, the response times for the congruent trials are subtracted from the response times for the incongruent trials to produce the IAT effect. To reintegrate; congruent trials are trials in which participants sort good words to white people and bad words to black people and in the incongruent trials the order is reversed. IAT effects vary from negative numbers (e.g., −200) to positive numbers (e.g., 300). For example, an IAT score of 300 would indicate that on average that person was 300 ms faster to sort the congruent compared to the incongruent trials.

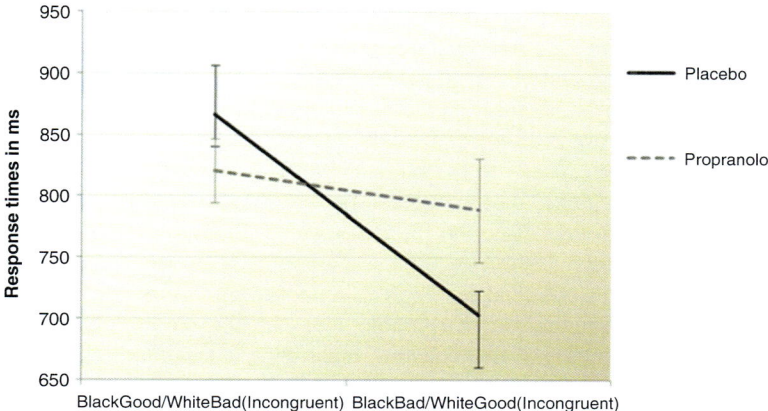

Fig. 4.1 IAT results, from Terbeck et al. (2012) (Reprint permission; open access article)

The explicit prejudice questionnaire is analysed by subtracting the answer to the questions, how warm one felt towards their own group compared to other groups. As previously mentioned, I am continually surprised that participants in the samples that I have used in my research never appear to score zero. Even though most people report that they feel as warm to white people as to black people, I never found a sample of participants where every single person reported this. Specifically, there was always someone, or more than one person, who explicitly reported that they felt warmer towards their own group than towards other groups. All data are then transferred to the final analysis software SPSS. Then, I believe comes the most important moment; inputting all of my data in SPSS, but there is one row empty, the row in which to write which participants took propranolol and which took placebo. At this stage the medical doctor would give the sealed envelope with the codes and to type this information in before completing the statistical tests. Firstly, we found no significant differences between the drug groups on the explicit measures, suggesting that explicit prejudice was not affected by the propranolol. Secondly, however, we did find significant differences on the IAT. Figure 4.1 shows the results of the IAT.

In Fig. 4.1, the x-axis represents the IAT condition (congruent and incongruent) and the Y-axis represents the response time in milliseconds. Further, the dotted line represents the propranolol group and the black line, the placebo group. Looking at the black line, the placebo group shows a strong IAT effect. Specifically, participants were faster in the congruent than in the incongruent condition. As also shown in Fig. 4.1, this effect was greatly reduced for the propranolol group. Besides this, more importantly, we found an interaction effect of task and intervention rather than a main effect of intervention using mixed ANOVA (a statistical test to determine differences in variances between groups). A main effect of intervention would mean that the group response times were generally different, but we did not find this. This suggests that participants in the propranolol group were not just tired or generally slower at responding. Indeed, we found an interaction effect of condition and intervention, which is also shown in figure 1. Participants in the propranolol groups were slightly faster in the incongruent condition and slower in the congruent condition, thus eliminating the IAT effect. Indeed, the results suggest that propranolol had an effect on implicit racial biases, but not on explicit prejudice. As discussed before, this suggests that noradrenergic mediated emotional arousal – fight or flight responses – might be causally relevant in generating implicit biases. This study thus produced the first evidence that a drug that is used to treat heart problems, and also anxiety, can also have an effect on higher order psychological processes, such as implicit racial biases.

Besides influences on racial biases, we found that propranolol also had another effect on a different higher order psychological concept; on an individual's moral judgment. Moral judgments are decisions about what is right and wrong, what one ought to do. For example, 'Killing is wrong'. We all agree? What about killing one to save five? Moral judgments were mostly originated in ancient philosophy. For example, Aristoteles would discuss cases of what should be done, and what ought to be the law. Later, psychologists have also investigated moral decision making, by examining how individuals arrive at certain judgments. An example of a classical moral dilemma (a situation in which moral rules might be conflicting) are the switch and the footbridge dilemma. In the switch dilemma, participants read a fictional description of a trolley that is out of control threatening to kill five workmen that are unable to escape the tracks and the approaching trolley. The reader should then imagine that there was a switch, which if pulled, would divert the trolley to an alternative track where there is one workman, killing him to save the five. Most people rate this action as morally acceptable. However, the rating changes when participant read about the footbridge dilemma. The scenario is the same; a trolley is approaching five workmen. However, now one should imagine that they were on a footbridge over the track and that there was a fat man on the bridge in front of them. They could now push the fat man on the tracks. His body would stop the train, and kill him but save the five. Now, most participants rate this as morally unacceptable, even though strictly speaking the outcome of the action is the same (one is killed, five are saved). It could be suggested that individuals judge the action in case two as less moral because it is a more "active" way of killing a person. One would have to actively push the man, rather than merely flipping a switch. Also, one might not believe that a body, never mind how fat, would stop a train (in fact it's actually not a train but a trolley). Or one might think about using the man as a means to save the five, while the other case feels more like the threat is being diverted from one party to the other. These and other examples of moral dilemmas were researched by Prof Joshua Greene in 2002, using fMRI. Greene et al. (2001) created a list of dilemmas, such as those described above, and distinguished between personal and impersonal situations. Personal situations, involve the footbridge dilemma, and impersonal, the switch dilemma. The authors found two key results. Firstly, that personal dilemmas, such as the footbridge dilemma, elicited activation in brain areas associated with emotion, whilst impersonal cases elicited activity in brain areas more associated with cognitive control and reasoning.

Secondly, they found that when participants made utilitarian choices in personal dilemmas (i.e., killing one to save many), the response time was enhanced and brain areas associated with cognitive control were activated. Greene, later developed the dual process theory of moral judgments, which states that deontological decisions (i.e., killing is always wrong regardless of the outcome) are driven by emotion and utilitarian decision are more driven by cognitive control. Indeed, deontology is a philosophical theory that proposes that some acts are intrinsically morally wrong (for example killing), regardless of the consequences. Thus, deciding to not kill one to save five could be called a deontological judgment. The other philosophical approach is related to utilitarian views, which postulates that the moral worth of actions is determined by its utility, the overall 'happiness', the consequences, or the best outcome for all. Thus, killing one to save five would be judged as morally acceptable. Greene proposed that emotional responses to the dilemmas elicit deontological decisions whilst further consideration and rational thought, overriding the initial emotional response, led to utilitarian judgments. There are now numerous researchers working on trolley problems; as well as creating alternative variations to the original scenarios. For example, would it matter which dilemma the subjects saw first? Or would it matter if there was a hole in the bridge and I could press a key to kill the fat man, rather than pushing him? Or what about if it was a fat woman? etc. etc. Recently a colleague of mine, David Edmonds, published a book entitled "Would you kill the fat man?" in which he describes the latest psychological and philosophical research on this topic. Here I want to demonstrate that higher order moral decision involving whether individuals judge an act to be morally acceptable, can also be affected by drugs. In our study, the participants again either took propranolol or a placebo before judging the moral acceptability of a number of impersonal and personal moral dilemmas, which included the footbridge and the switch dilemma. Participants then rated the degree to which they judged the action as morally acceptable, from not at all to completely. Besides this question, participants also answered the question, if they themselves would perform such an action. We found that propranolol did not affect the response to the question, of whether they would do it themselves, but it changed the response to the question if they judged the action as morally acceptable. But this only applied to personal dilemmas. Specifically, we found that participants in the propranolol group judged the action in personal dilemmas as *less* morally acceptable. At first glance these results might seem surprising, also since they do not appear to support Greene's dual process theory. Recall that Greene proposed that deontological judgments are more driven by emotion, and that propranolol reduces emotional arousal. This would suggest that participants in the propranolol group should have made more utilitarian decisions; but they did not. We could only speculate about possible causes for this effect.

One idea was that propranolol might have increased harm aversion by reducing aggression as supported by previous studies providing evidence for such a phenomenon. Furthermore, certain forms of aggression and readiness to inflict harm might be required to make the utilitarian choice, and this might be particularly relevant in personal moral dilemmas, such as the footbridge dilemma. A similar study was conducted at Cambridge University by researcher Dr. Molly Crockett. She used

SSRI drugs to modify serotonin neural transmission. Crockett et al. (2010) found that utilitarian decisions were associated with increased levels of serotonin, arguing that serotonergic transmission was involved in harm aversion. One usual comment when discussing moral dilemma experiments is that moral dilemmas are too artificial. I have often heard the comment:" What a train is out of control, and a fat man should stop it, how likely is that to happen in real life?" Thus, if psychologists are investigating how individuals arrive at moral decisions, it might be better to use more indirect or implicit methods.

Another problem is that people might say they would do a certain action, but if they were faced with that decision in real life, they would do something completely different. One famous example of this is the study conducted by Milgram. Milgram tested obedience, by asking subjects to inflict electrical shocks to another participant. In fact, the shocks were not real and the participant was a confederate of the experimenter, but the participants thought they were applying real electrical shocks with increasing intensity to another person. Every time this person got an answer wrong the experimenter asked them to give electric shocks to this person, starting with mild shocks (15 volts) leading up to painful and lethal shocks (450 volts). The confederate participant pretended to feel more and more pain. Despite hearing the participant in pain, Milgram observed that a large proportion of subjects obeyed the experimenter and continued to administer further shocks. However, when Milgram asked different subjects to imagine such a scenario, judging if they would obey the order and give electrical shocks, most people responded that they would not, even though most actually did in the real life situation. We recently developed a 3D interactive virtual reality version of the footbridge dilemma. The participant, wearing the Oculus Rift (3D virtual reality headset), finds himself on a footbridge with a fat man next to them. The train is approaching and a voice is heard saying:" Hey, I am too far away, but if you want to save those five people on the tracks you could push the fat man next to you on the tracks and derail the train. If you want to do it, do it now, but it's your choice." What we found was that most people did push him and moreover, we found that those high on psychopathic traits were more likely to push him than those with low psychopathic traits, whilst psychopathic traits did not predict the decision they made in the paper and pencil dilemma.

Going back to the pharmacological studies, the previous discussion has demonstrated that neurotransmitters are not only involved in physiological processes, or that drugs only effect the body, and possibly emotion, but that they can also modify, what one might call the 'core of a person', such as their implicit racial biases or their judgments of what is right or wrong. These studies are indeed not the only ones to show this, many more experiments have shown how drugs can impact on personality on decision making, on love, and on fairness judgments. We recently wrote a paper entitled "Are you morally modified", in which we describe the impact of a variety of drugs on such function.

One further well researched neuromodulator affecting on an individual's moral actions as well as on their emotion and their attitude biases is oxytocin. Many readers may have heard about the "love hormone", involved in maternal care, affection, etc. High levels of oxytocin can be recorded in woman before and after birth, and oxytocin is also medically prescribed to induce labour in pregnancy. In experimental studies, an oxytocin spray, or placebo spray is administrated to participants, and effects on psychological tasks are recorded. For example, regarding moral judgments, researchers have found that participants are more likely to donate money to a charity or judge offers as more fair with oxytocin than with placebo. De Dreu et al. (2011) investigated the effect of Oxytocin on racial biases and they found that, even though the IAT effect overall was not different, analysing the data in a different manner revealed that oxytocin increased in-group favouritism. Participants were more likely to associate positive words with one's in-group. This suggests that even though Oxytocin might increase love and care, but maybe it does so only for one's own group.

A further drug that has become more and more prevalent in research is Ritalin, or drugs containing amphetamines, due to their effect of enhancing some cognitive performances. In non-ADHD adults, amphetamines have been shown to increase working memory, enabling people to remember facts more efficiently. However, it was recently shown that this might be at the expense of making people less creative. Coming back to the main study of propranolol and implicit racial biases, a further recent study that we conducted confirmed our previous results, and also extended the discussion; in this study we were conducting pharmacological fMRI. Pharmacological fMRI involves participants receiving a pharmacological intervention (e.g., propranolol or placebo) while also taking part in an fMRI scan. The study design was quite similar to the first propranolol study. Participants were also screened for eligibility, they took the drug and waited for 1 h, and they also completed the IAT and the explicit prejudice questionnaire. Before they completed the IAT and the explicit prejudice questionnaire however they had a full brain scan.

Participants arrived at the Oxford Centre for Magnetic Resonance Imaging where they were given either propranolol or a placebo and were asked to wait in the waiting room. Heart rate was also measured every 30 min. Afterwards they were brought to the scanner room. As discussed before, the fMRI is a large magnet, so it is essential that participants are not carrying any metal inside or outside their body, as this could become attracted to the magnet resulting in physical harm to the participant. So besides confirming that participants did not have any mental implants inside their body, they also had to remove all coins, cards, jewellery etc. They were then asked to lie on the scanner bed which is inside a large tube with the mental coils around them. They could however always be released, by pressing an emergency key, in which case the scanner technician would immediately come to help them out if they felt unwell or wanted to stop the scan for any reason.

The fMRI expert was outside of the scanning room, operating the scanner and thus was on hand to help if any difficulties were encountered during the procedure. They could be in constant contact with the participant via an intercom, over which they also gave them instructions for the tasks. The first scan was an anatomical scan, as this was needed for the later data analysis, in which the anatomical brain picture for each participant was combined with a standard brain model. In the experimental task, participants viewed black and white unfamiliar faces in blocks. The participant's task was to judge the age of the person depicted in the picture. This task (judging the age) was merely there to keep their attention on the pictures and to ensure that they were indeed looking at the pictures carefully and were not consciously assessing the race of the person.

The task itself was quite short – lasting only about 15 min. After this, participants completed a final task in which they saw a purple and yellow checkerboard. This final task was a control task to ensure that the drug had no effect on overall blood flow because such as effect could invalidate the fMRI analysis. Brain activity to the checkerboard in the visual cortex was analysed, and it was confirmed that there were no differences in general blood flow between propranolol and placebo. After the scan, participants completed the IAT and the questionnaires, and finally were paid and thanked for taking part in the study. The results of this study confirmed those of the first study in that we also found a lower IAT score, but no effect on the explicit prejudice measure with propranolol compared to placebo. For the fMRI task, we found activity differences in the fusiform gyrus and in the thalamus. In particular, we found this effect for the contrast black face > white face, & placebo > propranolol. This meant that propranolol reduced the activity in the fusiform gyrus and in the thalamus for black faces only. In addition, to this we also conducted a time course analysis, that is, we further decomposed the fusiform gyrus activity to black faces and white faces over time during one block. With placebo, when viewing a white face, over time the activity in the fusiform gyrus appeared to decrease. The same pattern was observed for white faces when taking propranolol. However, for black faces activity over time increased with placebo, and this effect was not found for participants taking propranolol. This finding that the fusiform gyrus activity declined over time to white faces but increased to black faces has been previously observed by other researchers. However, we found that propranolol reduced the increase over time to black faces, suggesting that propranolol affected on the pattern of fusiform gyrus activity for black faces. Our results suggest that indeed noradrenaline – as mediated by propranolol - has an effect on the brain activity to other race faces.We did not find, as we might have expected though, differences in amygdala activity. The amygdala is the brain area strongly associated with emotional arousal, and it would thus, have been expected that propranolol might have an effect on emotion processing and thus on the emotion that was associated with the faces. However, what we found was that noradrenaline affected how the faces were processed initially. It might be suggested that noradrenaline was involved in the processing of other race faces, in that changes in basic emotional arousal, and reduced levels of noradrenaline, could have changed how they "saw" the face. For instance, Caucasian participants showed sensitization of the fusiform gyrus activity to black faces over time; propranolol reduced this sensitization, making the perception of white and black faces over time more similar. However, we also did not find that the activity in the fusiform gyrus correlated with the IAT, thus, we cannot say that this change specifically generated the reduction in the IAT score.

To summarize, in this chapter I described how the drug propranolol has an effect on behavioural measure of implicit racial biases, as well as face processing of in- and out-group faces in the brain. Furthermore, studies showing that psychopharmacological interventions can also have an effect on further higher order processes, such as moral judgments were also discussed. The results of these studies lead to several approaches; firstly, it should have become clear that drugs cannot only have an effect on an individual's physical state, but that drugs can also have an effect on higher order psychological processes, such as one's attitudes or moral judgments. Secondly, the results suggest that noradrenaline plays a role in generating the implicit racial bias. And thirdly, noradrenaline has been found to have an effect on the processing of other race faces in the brain. One can only speculate that maybe fear, or aggression, or both, might have contributed to this.

Open Questions Chapter 4
- Researching intergroup relation using psychopharmacological methods; what would you have done differently?
- Do you think that giving Propranolol to an extreme racist would change their attitude?
- Do you think neuroscience intervention have stronger effects than social interventions?
- Can you think of follow up studies based on the ones described in this chapter?

References

Crockett, M. J., Clark, L., Hauser, M. D., & Robbins, T. W. (2010). Serotonin selectively influences moral judgment and behavior through effects on harm aversion. *Proceedings of the National Academy of Sciences, 107*(40), 17433–17438.

De Dreu, C. K., Greer, L. L., Van Kleef, G. A., Shalvi, S., & Handgraaf, M. J. (2011). Oxytocin promotes human ethnocentrism. *Proceedings of the National Academy of Sciences, 108*(4), 1262–1266.

Greene, J. D., Sommerville, R. B., Nystrom, L. E., Darley, J. M., & Cohen, J. D. (2001). An fMRI investigation of emotional engagement in moral judgment. *Science, 293*(5537), 2105–2108.

Terbeck, S., Kahane, G., McTavish, S., Savulescu, J., Cowen, P. J., & Hewstone, M. (2012). Propranolol reduces implicit negative racial bias. *Psychopharmacology, 222*(3), 419–424.

Chapter 5
Neuroethics of Social Enhancement

In this chapter I will discuss the following questions; (a) Is prejudice in our brain? and (b) Should we use drugs to change ourselves? Let me first turn to (a) Is prejudice in our brain? I am not sure why I think this question needs discussion at all. But let me give you this example of a conversation between another researcher and myself at a conference in London. I told this person about my work and received the reply: "I am really quite angry about your work. Because this would suggest that racism was in the brain!" After a moment of shock and disbelief I replied: "As opposed to where?" Furthermore, news coverage of the propranolol and prejudice study included the report of a psychiatrist stating that a pill could not help racism as it was a learned behaviour. My reply would be: "Yes and where do you think learned behaviours are based?" Additionally, recently, in the Quarterly Review, James Watson (also author of 'The Psychotic Left'), published an article discussing the propranolol study, which he entitled "Implicit Racism as a medical condition". The author suggested that the research might stem from a strong left-wing attitude, which had motivated me to think that racism was an abnormal medical condition. First, and foremost, I am a scientist and my research is never politically or otherwise motivated at all. My research is only motivated by the scientific will to understand human nature. Secondly, the author wrote that the study presumed that 'race' has a biological basis. My reply would be:" Yes, everything has a biological basis." Indeed this also highlights a possible misunderstanding of our current understanding of the brain. To say it again, "*everything* has a biological basis"; our personality, our morality, and indeed our prejudice. This does not mean that it is necessarily inherited. Regardless of where it comes from and what it is, it is in our brain. Thus, interfering with the functioning of the brain can change your core of beliefs.

Historically, many people believed that there was a soul outside the body. For example, Levine (2010) described Plato's (427–347 BC) philosophy that illustrated a body and soul dualism, suggesting that the soul was immortal, and that the soul was imprisoned in our body. Damasio, a neuroscientist investigating emotion and decision making, entitled his book 'Descarte's Error'. The error was that the seventeenth century philosopher, Rene Descartes, believed that the soul was separate

© Springer International Publishing Switzerland 2016
S. Terbeck, *The Social Neuroscience of Intergroup Relations: Prejudice, can we cure it?*, DOI 10.1007/978-3-319-46338-4_5

from the body. In 1641, Descartes published Mediations on First Philosophy, which included his theory of mind-brain dualism, suggesting an immaterial soul distinct from a physical body. Later, philosophers also referred to "Physica Subterranean, when describing an invisible entity that controls the core personality. Also there are descriptions in the literature of mind-body dualism. He woke up and was a beetle; that's what most people remember about Franz Kafka's classical novel "Metamorphosis" Kafka (2001). In fact, the novel further describes the life of Gregor Samsa as a beetle, and the responses of his family to his transformation, and how he died in the end. Central to this novel is the fact that Kafka had the body of an animal – but that his mind was human. Indeed the novel starts with Gregor waking up and then realizing that he has the body of a beetle. He then wondered about this, wondering how he would appear at work with such a body, wondering how he could get dressed, and get on the train with an animal body. This lets us presume that Gregor was *not* really a beetle; as if he was he would not be wondering about such things. Indeed, the author might have considered the idea of mind-and-body dualism in describing a story involving the possibility that one has a human mind in an animal body.

Another novel, in which the author also divides body and soul, is Peter Dickinson story 'Eva' Dickinson (1988). In this science fiction novel, Eva's brain remained intact after an accident that destroyed her body, and was implanted into another body; into the body of Kelly, a young female chimpanzee. Eva thought: "The thing is you aren't just a lot of complicated molecules bundles together inside a skin – you're that too, but that's not what makes you *you*. ... Perhaps it isn't my dream at all – not Eva's, I mean. Perhaps its Kelly's...The thing is you aren't a mind in a body, you're a mind and a body, and they are both you.' Thus, the question regarding "prejudice" being in your brain seems less clear than I previously thought. It might be that people have the idea that one's constitutional factors, or maybe even ones basic emotions, are related to brain functions, but somehow when it comes down to morality, prejudice, and personality, then people might believe that these higher order, core functions are somewhere else. Or in the heart? As discussed in Chap. 3, even though neuroscience is not currently perfect, research though has determined that all human functions, from basic perception, to pain, to prejudice and morality, have a basis in the brain (and in the body). Furthermore, it is difficult to distinguish between concepts. For instance it might be easier to believe that one's feeling of pain was in one's brain, than one's feeling about other groups, but as we understand complex concepts more and more we might also be able to break them down to more basic principles. For example the prejudice as we found, might involve basic emotions such as fear or aggression, so taking away these basic emotions, might change behaviour and attitudes as well.

As discussed in Chap. 3, it was found, for example, that animals and humans with selective amygdala brain lesions showed very different behaviour, for example they were more ready to approach others, not shy, very open, but also not cautious at all. Another famous example that is often mentioned is the case of Phineas Gage, whose brain damage to the ventro-medial prefrontal cortex profoundly (and selectively) changed his personality. He became impulsive, suddenly begun partying, gambling, drinking whilst showing no impairment in IQ, or on other cognitive functions. Brain lesion studies demonstrate most clearly that changes or damage to the brain can have selective effects on personality and on attitudes. Thus, it should be obvious to conclude that all aspect of a person have a basis in the brain- our pain, our love, our prejudices, and our morality. We are our brain. However, this does not need to be confused with the idea that all those concepts are inherited, or that all those concepts are unchangeable. Here, to answer this question, it is indeed irrelevant if a function is inherited or learned, or both, it will still have a basis in the brain.

Take the example of riding a bike; this is surely learned. One is not born with the ability to ride a bike. One might be born with an ability to control motor movement, to coordinate eye and movement etc., but this particular function *to ride a bike*, is learned. Also imagine that someone would have a terrible accident, which involved damage to the cerebellum; this might selectively impair the ability to ride a bike (and certain other well learned functions) whilst the ability to learn new motor movements is still intact. Such an example would also support the fact, that regardless of how behaviour is developed, it is still correlated to changes, or neural activity, in the brain. Furthermore, another study found that taxi drivers had enlarged hippocampi (brain areas associated with memory).after learning road names and directions

Another famous example is the case of Patient H.M, intensively investigated by psychologist Brenda Millner (who is, by the way, over 80 years old now and still conducting research). Patient H.M. had a selective brain lesion, as he had – at that time – part of both hippocampi surgically removed to relieve severe epilepsy. H.M. was able to recall memories from his childhood, he was also able to recall memories involving well-learned motor actions (such as riding a bike) but H.M. was not able to form new memories. For example Prof. Millner gave him the task of drawing the lines of a star, whilst only looking at the template star in a mirror. This is a quite complicated task, and it involves learning new motor skills. But everyone – also healthy controls – struggle with the task to start with but then learns those skills over time in a number of different sessions. H.M. also learned the task. However, every time he came to a session, he had forgotten that he had done the task before. So even though he did not know he had done this task before, he was still better at it.

This demonstrates that the ability to learn motor skills (motor memory) is independent from the memory to store new factual memories. But beyond this, this example also demonstrates that functions, such as memory, motor control, and as discussed before, personality and behaviour are associated with the brain. "So you may think that its 'you' who likes the blue shirt or pink blouse with the dark jacket – but if our genes are playing a major role here, what do we mean by 'you' or 'me'? Maybe it's truer to say that we are experiencing the choice of our brain. Strange isn't it?" (Gordon 2005, p. 63). "People often wonder if there is a 'real self'… However, this is partly an illusion. There is no 'real-self'; there are only states of mind and patterns of consciousness. There is no single atom in your body that you were born with; in fact you are constantly changing the physical fabric of your body all the time, including your brain." (Gordon 2005, p. 106). Now, we (hopefully) agree we are our brain, let me describe some examples of how the environment, learning, reading, etc. etc., can all also change your brain, without the use of drugs.

At the 2015 European congress of Psychology in Milano keynote speaker Prof. Lamberto Maffeti in his talk "New frontiers in neuroscience: the environment and the brain", discussed how environmental influences can have profound effects on the brain – similar to those observed with drugs. Indeed, one key aspect of the brain is its plasticity. The human brain has the life-long ability to transform its neuronal connections based on experiences. Prof. Ungerleider, chief of the laboratory of brain and cognition at the USA NIH examined the effect of brain plasticity while subjects were learning new movements. Over a course of 3–4 weeks subjects learned sequences of new finger movements, which led to changes in the activity patterns in the prefrontal cortex as well as the primary and supplementary motor cortex. Most importantly the researchers could observe those changes in the brain one year later, even in the absence of further training. This goes back to the fact that everything changes your brain; in some way it could thus even be argued that reading a book was like taking a drug ; well, they are similar in terms of both changing your brain. Let me give further examples; Milton Erickson referred to hypnosis in which something is indirectly suggested to the patient;"I am going to hypnotize you. Please close your eyes and relax". His suggestion could then lead to the patient changing their behaviour, and to changing their brain. The influence of environment on development of stereotypical thoughts might also be demonstrated by apparently "innocent" child games. In many western countries, at least up to 2010, it was part of accepted the sports and teaching curriculum to play a game in which children catch each other, and this game was entitled "Who is afraid of the black man?". Only recently has multiculturalism had the effect that people abandoned this offensive terminology. Another example, which may be a more extreme way to change the brain with other means than taking a drug, is what one might call "brainwashing". On the 3rd of March 2015 the newspaper Daily Mail reported a story in which a mother claimed that her children were being "brainwashing into becoming racist". Apparently, her ex-husband, a strongly right-wing oriented person, had talked their children into having this attitude.

Even though one might wonder if this could have happened, in her book about "Brainwashing and cults", Singer (2003) starts with the sentence that she had always been interested in words "how words create mental pictures, how those pictures stir emotions, and call up other ideas and feelings, and how people use words to entertain, educate, and influence once another." (page XXI). Subsequently, Singer (2003) described how cults use 'persuasive techniques' including words. She calls this organised psychological and social persuasion that causes the members of the cults to change their attitudes, to do what the leaders wants them to do, and wants them to think. For example one former cult member described how he saw that he could escape this 'brainwashing' in theory but how it felt impossible for him to do so:" … I suppose I could have walked out of the apartment and away from it all, but I didn't. It simply never occurred to me." Singer (2003) further described how a long term process of brainwashing can slowly lead to an individual changing and be: "…suddenly confronted with a new physique". She stated that what was significant and wrong about these persuasive techniques was that people were of course not asked for informed consent to have their attitudes – and indeed their whole life – changed. And secondly they also, Singer writes, were not informed about side-effects, such that their life would be changed for the worse. Singer (2003) described many of such 'persuasion techniques' in her book. She discussed how one can manipulate people using trance and hypnosis, trickery, personal history revision, emotional manipulation, and peer pressure.

Besides such extreme "techniques", there are also normal, everyday, activities that can change your mood, change your attitudes, and change your life. Indeed, reading a book might be one candidate. Also, poetry, yes poetry, was also named as being associated with so called "sublime ecstasy". Romantic poetry, for instance by Thomas De Quincey, was suggested to be so strongly able to induce images in ones' head, and triggering imagination that this might have produced a s trance-like state. Another new technique that might involve changes in your brain, but not by using a drug, is mindfulness. Mindfulness is a form of meditation in which the person focuses on the present moment. Mindfulness has now been endorsed by numerous celebrities, suggesting that it reduces depressive moods and anxiety. Indeed, a recent pilot study suggested that it might also reduce racial basis. However, in a recent book entitled "The Buddha pill: Can Meditation Change You?" the authors argue that for some people meditation might have also negative "side-effects". Furthermore, psychological treatment for example cognitive behavioural therapy is intended to change a person, change their thinking pattern and their attitudes, and therewith also their brain.Cognitive Behavioural Therapy or CBT is currently the most used form of psychological treatment. It is based on the assumption that certain mental health problems result from dysfunctional thinking (cognitive) patterns, and from habitual behaviours, so that education, support and various techniques can help to reduce the patient's symptoms. For example in cases of anxiety CBT has been shown to be very effective, especially when treating a variety of phobias. Here a combination of (un-)conditioning as well as cognitive intervention is used. Take the example of fear of spiders; Mr Brown is very afraid of spiders, he can't see them, can't touch them, and screams out if he is close to them. He has now started to avoid going into the cellar in case he encounters a spider, also he feels very uneasy when trying to sleep unless he checks all walls for spiders before. His fear has thus taken over his social life and he has therefore developed into a phobia for which he is seeking help. He is going to a CBT based psychologist who persuades him to slowly overcome his fear by taking gradual steps to increase his exposure to spiders over several sessions. . First perhaps, he might be able to be in the same room with a spider, then looking at it directly and focusing on it, then finally touching it. Besides this, CBT assumes that he has some dysfunctional beliefs regarding his response to spiders. For example he might think that he would faint, or have a heart attack because of his anxiety. Thus the psychologists talks to him, tells him that his anxiety might be strong at first, but that it will pass. Also the psychologist will tell him that a fast heart beat and other physiological responses are part of a normal reaction and that the worse that can happen is just that he feels uneasy at first, but better after a while. This is just a brief example of course, but this illustrates the following; a combination of conditioning and talking can change someone's behaviour, their thinking, their brain.

I recently found an article in the BBC news with the following headline; "Sleep training may reduce racism and sexism"[1]. The original study was published in the journal Science, and the researchers, from North-western University in Illinois and Princeton, had participants processing counter-stereotypical information which was paired with a special sound Hu et al. (2015). For example in the lab participants saw a female face and the word "math" on the screen. They were then instructed to respond to this (via a key-press), but to not respond when they saw a picture of a woman and for example the word "housework". So the participants learned to only press the key when they saw counter-stereotypical information. Correct responses were then followed by a sound. The training worked, in that after completing 360 trials, as described above, the participants then completed an IAT again, and showed reduced bias. This is not surprising, as they had just finished completing a large number of trials training them to respond in a counter-stereotypical manner. However subsequently, the participants were asked to take a nap and their EEG was recorded. When it was shown in the EEG that they were in slow-wave sleep state, the sound from that task (that always appeared when counter-stereotypical information was presented) was played. The researchers then found that this sleep training fortified the reduction on the IAT compared to a group that did not have such training, and that the reducing effect remained even 1 week later. This suggests that reactivation of the counter-bias training during sleep increased the effectiveness (in reducing biases on the IAT). Well, that certainly might remind individuals of 'Brave New World' scenarios, in which the novel describes how individuals in the future get suggestive messages played to them during their sleep. Indeed, in a BBC news article Gordon Fell & Jan Born, researchers from the University of Tubingen, and commenters on the paper, and, cautioned that sleep was a vulnerable state, in which people would not have wilful consciousness. In Brave New World – the English classic science fiction novel from Vintage Huxley (2007) – a dystopian future is described, a world in which everyone should be happy. But at what price? In the novel everybody is taking the pill *'soma'* that delivers instance bliss with no side-effects. Besides taking soma, people are also genetically selected to have the best outcome. And there is another force; conditioning. As babies, people hear the same messages over and over. For example imagine being selected to be a hair-dresser, so from a very young age, make up, hair products etc. get shown to you and while this happens, positive things happen as well, for example you get shown nice pictures, hear good music or get nice food. However when shown other objects, such as books or tools, you get extremly loud noises played or electric shocks. Also, to be found in this novel, intergroup conflict might have been conditioned. Consider the following abstract from Brave New World; " 'Elementary Class Conciousness, did you say? Let's have it repeated a little louder by the trumpet.' At the end of the room a loud-speaker projected from the wall. The Director walked up to it and pressed a switch. '…all wear green,' said a soft but very distant voice, beginning in the the middle of a sentence, 'and Delta children wear khaki. Oh no, I don't want to play

---

[1] http://www.bbc.co.uk/news/health-32914228

with the Delta children. And Epsilons are worse. They're too stupid to be able to read or write. Besides they wear black, which is such a beastly colour. I am so glad I am Beta.' There was a pause; then the voice began again. 'Alpha children wear grey. They work much harder then we do because they are so frightfully clever. I'm really awfully glad I am Beta because I don't work so hard. And we are much better than the Gammas and Deltas. Gammas are stupid. They all wear green, and Delta children wear Khaki. Oh no, I don't want to play with Delta children. And Epsilons are still worse. They're too stupid to be able to …'The director pushed back the switch. The voice was silent. Only its thin sound continued to mutter from beneath the eighty pillows. 'They;" have that repeated forty or fifty times more before they wake; then again on Thursday, and again on Saturday." …. "'Till at last the child's mind is the suggestion, and the sum of suggestions is the child's mind." This is a fictional scenario, but it should have become clear, that everything can affect the brain, to a lesser or stronger degree, and that 'other things' can be as powerful as drugs. For example, Mary Gordon, involved in child psychology, and child moral education stated that:"Empathy is integral to solving conflict in the family, school-yard, boardroom, and war room. The ability to take the perspective of another person, to identify commonalities through our shared feelings, is **the best peace pill that we will have**."

5.1 What Is So Bad or Different About Drugs?

In 2015, in the UK city of Exeter, I presented my work, when I met another researcher, who I believed was concerned, unhappy, and offended, by the idea that one would take a drug to affect racial biases. She told me about a film entitled "Equilibrium", a science fiction movie about a future time when no emotions were allowed. Equilibrium is a 2002 American science fiction film produced by Kurt Wimmer. The action is set in 2072, when the survivors of a World War III live under a totalitarian regime that forbids any emotions because the government determined that emotions are the only causes of prejudice, violence and war. Indeed, someone who had emotions would be killed. The state ordered everybody to do two things – and this is important – to (a) take a drug that suppresses emotion and (b) to eliminate all emotion eliciting material such as books, pictures, and music. Again, we see here that even in fiction the writers seem to address the issue that a drug might not be sufficient and that other factors, such as books, music etc., can also have profound effects. Indeed, in the previous section it was shown how everything can have an effect of the brain. So what is so bad about drugs? In the remainder of this chapter I will address this question. As already discussed in Chap. 3, in addition to proprano-lol, there are many other pharmaceuticals that have been shown to produce effects on personality and morality. Furthermore, besides drugs affecting higher order functions, there are also many drugs that are used not only to "cure" medical or psychiatric problems but to "enhance" *normal* human functioning. First, it should be considered, "what is normal", and when does something count as treatment and when does something count as enhancement. For instance if someone is suffering from severe depression and receives a drug most would agree that this is treatment, but what if someone is slightly unhappy but wants to just feel a little better and takes a drug, is that enhancement? And coming to this, who determines the cut-off? This is an important question, as to date many people talk about what they refer to as 'medicalization of society'. For example, mentioned that he had many patients in his clinic who wanted to use drugs to overcome difficult times, such as coping with the death of a loved one. He was questioning the idea if it was desirable to "fight against perfectly normal human emotions and feelings." (p. 150). Therefore we can ask the question; who determines the cut-off? When is something a problem and when is it not? Or indeed, is it necessary to enhance? Or, is it even possible?

The problem of psychological or psychiatric diagnosis is indeed important and might be misunderstood. For instance one could go to their GP, have their blood examined and it can be 100 % determined if one has for example diabetes. With some medical condition, there are clear factual markers, and the diagnosis is based on that. With mental disorders this is not possible. Indeed, even in cases of schizophrenia, or autism etc., there is no one 100 % test to determine this. Thus psychiatric diagnosis is much more subjective compared to medical diagnoses. For example if a patient came to see their doctor complaining of "being depressed", how will this further be determined? Specifically, how is it determined if he is suffering from a depressive mental disorder. And furthermore, if he is not, does he still need treatment? Firstly, let me address the question, how the diagnosis is made, before discussing treatment. Both are very difficult questions and of course there are no easy answers. The doctor, when trying to make the best diagnosis, would write down the symptoms that the patient reports, they would also give them questionnaires about their symptoms and if available perform psychological tests, and if possible obtain information from collateral sources, such as family or friends etc. If all this information together supports the diagnosis they might come to this conclusion. In addition, as part of the diagnosis, other causes for low mood etc., for example medical illnesses, have to be excluded. One common factor which is usually taken into account is impairment within one or more areas of life. For example, is the problem affecting the person's ability to work or to socialise? However, even though a clear diagnosis might be important in some cases, for example if someone requests to receive early retirement or disability benefits, this does not imply that someone who is seeking help can just be told: "You have nothing. Go away." As even though they might not have a problem which fits the full criteria they still are there and feel they have a problem, and should thus receive help. But should the help take the form of drugs? And if it does, does this count as enhancement? There are many cases, in which the term enhancement might be easier to follow. We generally distinguish between physical, cognitive, emotional, and moral enhancement. The term enhancement however does not imply that a drug is involved. Indeed, I will later discuss in this chapter how moral education (for example in school) could also be termed moral enhancement. And also training in sports can be termed enhancement. The cases which I will however first discuss here are cases of pharmacological enhancement.

In his book, Prof David Nutt presented data that supported his previous controversial comment that "horse riding was more dangerous than taking ecstasy". A recent article in the Washington post for example suggested that LSD was found to make people smarter and happier, and helped alcoholics to drink less. He also discussed cases in which the line between medical and recreational use of drugs was not clear cut. For example some patients also use their prescribed drugs without following strictly medical directions. Prof. Nutt also discussed cases in which a "forced" intervention was very helpful to the patients. Indeed, forced feeding in severe cases of anorexia nervosa can save lives, as can medication in cases of severe depression involving a risk of suicide. Such patients do then subsequently mostly report that they have been pleased about the life-saving treatment that they received, even if they might not have wanted it at the time when they were ill. However, more recently we also find more and more references to cognitive enhancement. Prof. Barbara Sahakian published a paper in the prestigious journal *Nature*, entitled "Professors little helper." Here she described the observation that more and more academics as well as students were taking drugs to enhance their academic performance at university. The most common drugs used were analogues of Amphetamines, which are either used as recreational drugs, but also as treatment for ADHD and narcolepsy. Examples of such drugs include Modafinil and Ritalin. Generally, the drugs have been shown to increase alertness, and are used by shift workers; drivers etc., to increase the time they could remain alert and awake. In academic situations, the drugs might be consumed in order to study (all night before the exam) without falling asleep. But besides the mere increase in alertness there is also some evidence that memory functions might be increased, and that therefore the students might remember more. However, recently it was determined this this might come at the cost of students being less creative. On a side note, obviously this implies that students must believe that university education is just about memorising facts, which is a misunderstanding, as even though the drug might lead to people learning more facts, university is not just about memorising facts without understanding them. Prof Nutt discussed cases of performance enhancement, stating that anabolic steroids are the most widely used drugs for physical enhancement. These drugs mimic the male sex hormone testosterone and stimulate growth and the "androgenic" part. Many drugs have also been used during war, and within battle zones, for example large quantities of morphine were used by soldiers in the Franco-Prussian and American Civil Wars. Besides this, amphetamines were thought to increase military superiority, for example they were used by German soldiers during the 2nd World war.

Prof Nutt also suggested that pharmaceutical treatments might be used to make psychological treatments more effective, for example as an additional aid to treat various phobias. Turning to emotional enhancement; obvious candidates might be SSRIs (for example citalopram) or benzodiazepines. As described before we would here need to consider people who would like to take this drug to feel extra happy or not anxious at all. By the way, currently this is not legal. For example I could not go to my GP and say I wanted to have a drug to "enhance myself". This is of course also the case in terms of performance enhancement, no GP would prescribe Ritalin to a student so they can memorise better or study the night before the exam. Thus the use of drugs for enhancement purposes have to be obtained illegally, which means for example that people might buy them from some (unknown) internet provider who sells the drugs without the required prescription. This is obviously potentially dangerous, and one might accidentally buy a dangerous or lethal substance, or one might simply buy a rather expensive placebo.

Back to the case of mood enhancement. There is in fact little evidence that there are drugs that enhance mood in healthy volunteers. Indeed in a recent large metaanalysis it was determined that for example anti-depressants only seem to show a significant effect in people with severe but not moderate or mild depression. Thus, there is no clear evidence that it is indeed possible to enhance mood in healthy volunteers. Finally, let us discuss moral enhancement; what is moral enhancement? Indeed, what is the optimal state one wants to achieve? I will discuss this in further details in the final chapter, but here I want to just briefly address the question that compared to others forms of enhancement the optimal state for moral enhancement might be less clear. Although this might also be the case for mood enhancement. For example if one takes a pill to be happier, is it clearly "better" to not be sad at all? The academic area, which investigates issues related to enhancement, is called Neuroethics, a sub-discipline of philosophy. The key philosopher and researcher in this area is my collaborator Prof. Julian Savulescu, who was the first to stimulate philosophical, scientific, as well as public debate about cases of enhancement. Later, in one article on enhancement other researchers suggested that reading a book was "on some level" similar to deep brain stimulation (an invasive procedure to electrically stimulate certain areas of the brain) as they both change the brain. When I first read the article, I was very surprised and in disbelief, as on the mere face these two things are obviously very different. In one I sit down open a book and read for a little while, in the other I go to hospital, have an anaesthetic and a major intervention. So what is this about then?

Indeed, some researchers have argued that as both (reading a book and deep brain stimulation) have an effect on the brain, they were similar; well, "morally similar". In the same manner, drugs and reading a book (training) could also be viewed as being similar. One argument against the claim that drugs and books are the same might be the suggestion that drugs are synthetic or "unnatural". This argument can however be easily addressed. For example what does the term "natural" mean? Does it mean it is not produced from artificial chemicals? So would drugs that are based on plants be ok then? What about opium then? Indeed, opioids are drugs that are either derived from the poppy plant, from opium and morphine themselves, or are synthetically created to act like opioid analgesics. In addition some "natural" mushrooms, 'magic mushrooms', are naturally produced but can have very strong effects on humans and can even lead to enduring mental illness. Furthermore, wearing clothes, eating processed food or even cooked food, wearing make-up, cream, shaving, etc. could all be described as "unnatural"? Therefore it is easy to argue that the difference between a drug and other interventions cannot be that a drug is not natural and the other things are, as this is simply not the case. The second argument against the idea that taking a drug and reading a book are similar is autonomy. Indeed, many people might believe that taking a drug must somehow involve force. Fear of drugs might also have come from stories about previous times in psychiatry when indeed patients were given a variety of drugs –including LSD – as part of experiments or for other purposes. Ronson (2011) furthermore described scenarios in which LSD drugs were given to CIA assassins to brainwash them.

A further problem might be overmedication and maybe also over diagnosing of mental illnesses, so stated Ian Goodyer, a professor of child psychiatry at Cambridge University (cited from Ronson 2011). Furthermore, it might be feared that drugs can cause a person to be in a state that they cannot control whilst if they are reading a book they can just put it down if they don't like it. First and foremost, it is of course the fact that no one is – or can be – forced to take a drug, just as they cannot be forced to read a book. Another example is the case I described before, a convicted paedophile, who has abused children and is now receiving treatment, which includes the strong suggestion to take medication to reduce their libido function, but they are not forced. When are people forced then? This might happen in cases of forced feeding in severe cases of anorexia, in which relatives give consent. Therefore scenarios of science fiction in which the government for example can force people to take drugs, seems very unreal. A government could just add drugs to the water supply, forcing people to take drugs without them even knowing. Again, this would require a totalitarian, corrupt government. However, even if one were to agree that people may not be forced to take a drug, they might still voluntarily take it and then be in some helpless unescapable state, which makes them do things they don't want to do and there is nothing they can do about this. This fear might stem again from movies or from experiences where one took a pharmaceutical and experienced negative effects – side effects – that they were not able to stop.

For example individuals who take party drugs, such as ecstasy or magic mushrooms, might report strong changes in sensations, hallucinations, and effects which could also be experienced as being unpleasant but which the person is unable to control until the drug effect is over. In fact alcohol also produces changes within the person (for example reduced inhibition) which a person might believe they are not able to control. This question is quite challenging and involves a consideration of free will and determinism, because one argument might be that there are few things that one could control, and that also other – non drug – interventions might produce effects outside a person's control. For example imagine the case that you are watching a horror film and you find it awful and it will give you a nightmare the same night. Well one might then think "Oh dear, I am never going to watch that film again." One can do the same with a drug of course. Thus, the effect of watching a film or reading a book might also be termed "uncontrollable" in that sense. But I understand the point that during the process a drug effect might be stronger, and whilst someone can just press a key to stop the horror movie, one may have to wait hours for the drug effect to reduce. So, even though this argument is partly persuasive there is another important argument, which I also recently discussed in a research paper that might be the most important one in the debate, which is the safety of a drug compared to other interventions. What are the side effects of reading a book? One might argue that you will have less time to do other things. The book might be upsetting. However, these side effects seem clearly less severe than side effects which certain drugs have, such as cardiac problems, liver dysfunction, and even death. Some researchers have argued that advances in science might produce drugs that have less – or negligible – side effects in the future.

However, in our paper, we argued that it is very unlikely that there will ever be a drug with no or negligible side effects. This is the case because of the architecture of our brain and the distribution of neurotransmitters within neurons, which are interconnected, and widely distributed, thus necessarily producing unwanted side effects, alongside the desired effects. In the paper we gave an example of a fictional scenario, in which one wanted to enhance a function, which superficially, might be coded by a relatively small and local group of neurons. Imagine one wanted to enhance or improve their visual ability to detect edges in images. Just to remind you, this is a fictional scenario that was chosen because neurons and receptors which code for edge detection in the human brain are well established and localised. So one might think of a pharmaceutical that would target the receptors in brain regions of visual perception that code for edge detection. We then illustrated that the very same receptors that are involved in edge detection in one brain region, are however involved in multiple other functions in many other brain regions, such as regulating heart rate, sleep, and auditory perception. Thus such a pharmaceutical would also produce numerous side effects. And now imagine the case of moral enhancement, where the function itself already involves a large network of interacting neurons, how can it be possible then to produce a drug that would just target this one function? It is very unlikely to be possible. There might be developments that one could inject a pharmaceutical into a certain brain region, and this would not be distributed further. However, this might reduce side effects when we talk about localised functions, but not when considering higher order human functions, that all involve a large interconnected network of brain activity such as morality. Thus, the decision to take any drug, not only for enhancement, but also for any medical or psychiatric disorder, is always a trade-off between the desired effect and side effects. Mostly, in medical and psychiatric cases, in which the drug relieves suffering, or is indeed the lifesaving intervention, the desired effects much outweigh the side effects. However, when discussing enhancing, would one really want to take a drug to study all night before an exam, and then feel terrible for 2–3 days, maybe having sleeping problems, loss of appetite, danger of heart failure etc.? Thus, the trade off in cases of enhancing might be that the positive effect of the drug does not outweigh the side effects. Indeed, personally, I would *never* take a drug to enhance myself, because of the risks. And furthermore, I strongly suggest to anyone who is thinking of taking a drug just to be better or faster or happier, *not* to do it, because it is dangerous. Indeed, if there was a drug that made me fly, and nothing else, I could just fly and there were 100 % no side-effects then I would take it. But, there is no such drug, and it is also very unlikely that there will ever be. Thus, the work on the psychopharmacology of morality has been conducted in order to understand the processes in the brain which are involved in morality, but not to suggest that one should take a drug to change them. There is more than one reason for this; (a) It is dangerous, (b) It is not possible (as we have seen in Chaps. 2, 3 and 4, morality and prejudice are very complex and (c) It might not always be desirable; but more on that in the next and final chapter.

Before turning to the last chapter, there is one topic which should finally be briefly discussed here, which is moral education. Discussions on moral education might start with one key question: "Are we born good or evil?", or on a different note, are we born prejudiced? As I hope I made clear from the beginning of the chapter, that a drug might interfere with prejudice has nothing at all to do with whether it is inborn or not. But are we born with a tendency to prefer our own group? Or are we born to like certain people and to fight others. Are we born with aggression? Or just love? Is it society that makes us good, but we are bad? Or are we born bad and then we receive moral education and become "good". In Chap. 2, I discussed this topic a little, but here I want to add research evidence that is relevant to consider. At Yale University a group of researchers have attempted to answer this question by conducting experiments with toddlers, testing their morality. In the study the baby observed a toy which was either behaving badly (i.e., being mean to another toy bear) or nicely (i.e., helping the other toy bear). The researchers found that when the child had to decide which toy they want, the majority of children decided for the good toy, and not the mean one. This might suggest that they understood concepts of helping and harming, and choosing the nicer character for themselves. However, children can also be mean. One example which was at the time reported widely in the news in America apparently shows early tendencies to prefer one's own group.

As described in Chap. 2, the experiment is called the doll test. In this test the child can choose between a black and a white doll. Children were asked which the nice doll was and which the mean doll, and which the ugly doll. White children, wanted to have the white doll, found it nicer, and less mean. Black children wanted to have the black doll. In another study at Yale, researchers also found that young children at certain ages, were not likely to share, but rather wanted to keep sweets and tokens for themselves only. Psychoanalyst Freud suggested that we might be born with two tendencies (he called them Thanatos and Eros) a love and a destruction instinct. This might give rise to the idea that humans are born with tendencies for good and bad, for aggression and selfishness, as well as care and kindness. But society can certainly shape and strongly develop a person, and society and culture might also contribute to the behaviour which people display, and society might determine which actions are seen as acceptable. One social factor is the mere establishment of laws which allow for the punishment of anti-social behaviour. However, we make the laws, and we develop morally. Are there some universal moral rules? It might be suggested that, over the centuries violence and brutality has declined. Whilst for instance a few centuries ago, women had no rights, and could lawfully be beaten by their husbands; this has changed. Also corporal punishment of children is now deemed unacceptable in most societies.

Furthermore, when looking at forms of punishment, in history humans being burned alive, and brutally tortured, violence occurs far less frequently nowadays. One aid to moral development over time, were surely not drugs, but moral education. Indeed, the power of education might be stronger than one might think, and stronger than any drug effect. Moral education is teaching morality (what is good and bad), and happens everywhere; at school, at home, in books, on TV, on the radio, at work, in society. One might think that the term moral education was related to teachers or parents telling a child about what is right and wrong (i.e.,: "Don't hit the other child."), but moral education goes far beyond this explicit teaching, and implicit moral education takes place all the time, and not just for children. Who are the characters presented on TV? In the past there were many reports that the bad guy, the criminal, was often a black person. This problem has now been recognised and people are hopefully trying to portray characters accurately. The important message here is that such TV presentations can lead viewers to observe what is apparently "normal", how society functions, and if this is representing an inaccurate picture, then this teaches inaccurate moral values. Furthermore, if children in the past observed apartheid, and adults behaving terribly towards people from other races, they implicitly absorb similar attitudes. A similar issue is related to gender; whilst many job interview questionnaires today not only ask about gender (male, female), one is also asked about whether they are transsexual, or have changed gender. This was unthinkable in the past.

The way society functions thus also implicitly shapes the morality of the individual. Thus, it might be a combination of our tendency to be good and bad, and the way moral education happens in society. But it is indeed very difficult to determine how changes are implemented. With regards to prejudice, instead of taking a drug, there are means which can reduce prejudice that can be even more powerful, such as moral education. For instance, government implementations of programs to increase equality and diversity can also shape moral values. The establishment of laws can also influence this process. One might argue that this only however reduces explicit prejudice so that people are simply not overtly reporting any prejudicial attitudes, but it has also been shown that intergroup contact, and intergroup friendship not only reduces adverse comments about other group members, but also increases empathy towards each other, and impacts on their moral education.

Open Questions Chapter 5
- Would you take a drug to enhance yourself?
- Do you think people should be encouraged enhance themselves (with whatever method) in order to enhance human morality?
- Do you think reading a book is the same as having deep brain stimulation?
- Is there any benefit to be sad sometimes (and not always happy)?

References

Dickinson, P. (1988). *Eva*. London: Corgi Freeway books.
Gordon, M. (2005). *Roots of empathy: Changing the world child by child*. Toronto: Thomas Allan.
Hu, X., Anthony, J. W., Creery, J. D., Vargas, I. M., Bodenhausen, G. V., & Paller, K. A. (2015). Unlearning implicit social biases during sleep. *Science, 348*, 1013–1015.

Huxley, V. (2007). *Brave new world*. London: Vintage, Random House.
Kafka, F. (2001). *Die Verwandlung*. Stuttgard: Reclam.
Levine, L. (2010). *I think, therefore I am*. London: Michael O'Mara books.
Ronson, J. (2011). *Them*. London: Picador.
Singer, M. T. (2003). *Cults in our minds*. New York: Wiley.

Chapter 6
What Should Be Done?

As men advances in civilisation, and small tribes are united into larger communities, the simplest reason would tell each individual that he ought to extend his social instincts and sympathise to all the members of the same nation, though personally unknown to him. This point being once reached, there is only an artificial barrier to prevent his sympathies extending to the men of all nations and all races. (Charles Darwin)

6.1 Should We Cure Prejudice?

To review; we discovered in Chaps. 2, 3 and 4 that we could *not* cure prejudice. Indeed, even though some experimental studies might have shown that certain drugs have an effect on racial bias; this is far from a cure. We have no fixed, 100 % treatment measure for prejudice; the brain and drug effects are complex; we do not fully understand the interaction of neurotransmitters, real life behaviour is different to lab situations; individual responses are different, etc. etc. In Chap. 5 we determined that we also could *not* cure prejudice with medication as drugs come with side effects, and the effects of enhancement might not outweigh the side effects of pharmaceuticals. Finally, I want to discuss the ethics of "curing all prejudices". Do we want to have no prejudice in society? The question here then is whether it is desirable, and "natural" to eliminate prejudice. Would it be good to reduce all negative emotions? For no one to be aggressive? For us to love just everyone? To have a world, where there is no fear or depression, and no prejudice?

Indeed, it seems that we have fear and aggression as part of human nature. Hobbes wrote in Leviathan (1651):"…So that in the nature of man, we find three principal causes of quarrel. First, competition, secondly, diffidence, thirdly glory. The first make the men invade for gain, the second for safety, and the third for reputation. …". But maybe all we need for a better world is to have no negative emotions. It was reported that Stephen Hawking, when receiving a prize for his life time achievements in science, mentioned in the ceremony, that aggression will be the

© Springer International Publishing Switzerland 2016
S. Terbeck, *The Social Neuroscience of Intergroup Relations: Prejudice,
can we cure it?*, DOI 10.1007/978-3-319-46338-4_6

downfall of the human race, and that all that was needed now (e.g., in contemporary society) was empathy. Indeed, there seems to be a very strong need to enhance empathy.

Krzanaric (2014) wrote about what he calls "collective empathy", a compassion that goes beyond the individual but recognises the wider society. He bases this argument on Steven Pinkers book "Better Angels of our Nature" and specifically to the humanitarian revolution in the eighteenth century. Krzanaric (2014) stated that especially today more empathy is needed; that we for example should care for the people who made the pillow when we wake up, and think about the people who provided the beans for our morning coffee. The power of empathy in reducing prejudice and discrimination was also recognised by de Waal (2010) "Empathy is the one weapon in the human repertoire that can rid us of the curse of xenophobia." Additionally, Gilbert (2010) stated that besides intensely cruel and callous behaviour, humans also show great capacities for compassion. He discussed how feeling loved and having friendships and care, significantly influences our own well-being to the positive. Thus, compassion he argued might be especially important in a contemporary competitive world.

Greene (2013) mentioned two central threats to the survival of humanity, one being natural disasters but number two being the ability to build weapons of mass destruction. Indeed, there might be evidence that enhancement of empathy might have contributed to enhancement of morality and reduction of violence and brutality. In his book, Steven Pinker (2011) argued that in ancient human history people were much more violent, and that civilisation moved humanity in a "more noble" direction. For example in ancient times there are references to a 'whipping boy', an innocent child who could be flogged in place of a misbehaving prince. Multiple example of extreme violence, face-to-face battles and torture, can be found in history. For instance in 800 BCE King Menelaus's brother described his plans for war:" Menelaus, my soft hearted brother, why are you so concerned for these men? Did the Trojans treat you as handsomely when they stayed in your place? No. We are not going to leave a single one of them alive, down to the babies in the mother's wombs – not even they must live. The whole people must be wiped out of existence, and none be left to think of them and shed a tear." The philosopher Peter Singer (1981) introduced the concept "The Expanding Circle" when describing that over history, humans have enlarged the groups of people they interact with, and share vales with, and have expanded feelings of empathy towards, with the closest inner circle being one's own children and family. Steven Pinker argued that the expansion of literacy might have contributed to this effect. Also other factors can have contributed to this such as the opportunity to travel world-wide, the internet, globalisation, immigration etc. As discussed in Chap. 5, law enforcement has contributed to increasing morality and equality. For instance advances have been made through government policies, making racist attacks illegal, and pursuing integration policies such as mixed schools, government, business, and education.

Steven Pinker stated that in the late 1950s only 5 % of white Americans approved of interracial marriage, which rose to 80 % in the 2008. This might demonstrate that social efforts have strongly contributed to a more equal and advanced civilisation.

Pinker called this the rights revolution, which enabled greater equality for racial minorities, woman, children, and gay people. Pinker stated that the immoral violence of the past has been replaced by a new way of ethics that is governed by empathy, rights, and reason. In addition, education has also helped people to realise that previous belief systems were often wrong and poisonous. There has been a need, for example, to overcome historical beliefs that children need to be beaten to be socialised, or that woman like to be raped, or that animals can't feel any pain. In America the civil rights movement was a further big step towards equality. Furthermore, in 1950s segregated schools were banned. But the efforts for equality are still not sufficient, as we still find many right wing groups in the USA, such as the Ku Klux Klan, inciting racist criminal acts. Only recently, in 2015, we heard in the news about a white racist who shot a number of black people in a community church. However, when talking about prejudice, people usually mention extreme past historical events, where overt extreme prejudice and racism caused terrible violence on a large scale. But would curing prejudice (if that was possible) prevent wars? It is well researched that besides individual's prejudice being involved in such events, we also have large contributions stemming from political systems, from laws and orders in society and from resources available to society etc. Thus, prejudice often might not exist in isolation but is also embedded within the social context. Krzanaric (2014) however also notes that often throughout history people believed that in order to create a good society one needed a gun in one's hands. However, even if there were no guns, there would be still fights, and I argue, that even if there was no prejudice there would still be wars. For example even in the the 2nd World War, in which antisemitism obviously played a huge role, Hitler also invaded Poland, Russia and other countries, motived by perceived limited resources, gaining of power etc. Steven Pinker explained violence, and moral justification of violence, partly through through ideology; the belief in an utopian future, and a utilitarian belief that genocide may be a means to it. He describes multiple causes of violence, such as ideology, sadism, dominance, and revenge. Haidt (2007) described five concerns that he called moral foundations, which may be in conflict; (A) In-group Loyalty, (B) Authority/Respect, (C) Fairness and Reciprocity, (D) Harm/Care (E) Purity/Sanctity. Thus, it becomes clear, that even though prejudice is one factor in conflict and violence, eliminating prejudice would not eliminate all violence, as people would probably fight for other reasons.

The 2nd idea might be that prejudice is a "disorder", something that needed to be cured. Indeed, in some articles, authors have suggested that racism should be classified as a mental disorder (e.g., Poussaint 1999), and the Oxford Handbook of Personality Disorders labels extreme racism as a 'pathological bias' (Widiger 2012). Furthermore, philosopher Blum (2004) suggested that: "false stereotypical beliefs can be bad even if they do not contribute harm to their target." and that attitudes, and not only actions, can be subject to moral evaluation. However, a sweet anecdote is this; Hartup (1979) quotes two poems by boys in a class of 9 to 10 year olds in an American elementary school. A white child wrote: "If I were black, I'd feel what black people feel. If I were black, I might be prejudiced against whites because whites would be prejudiced to me. It feels like being shot when someone is prejudiced to you. If someone hit me because they were prejudiced. My heart would be stung, like being stung by a bee." And by a black child wrote: "Black is black White is white. So why does Black give you a fright? I am black. You are white; to me black is a great delight. Some people say prejudiced people are bad. That is not true. They are not really bad. They just should not believe the way they do." There is a novel called 'Strolling Players' by Charlotte Yonge and Christabel Coleridge, published 1901. At one point the characters are talking about attitudes to class, the theatre, etc.: "Good people *are* prejudiced', said Juliet petulantly. 'I don't think they are much more so than bad ones', said Clarence candidly." A recent BBC documentary showed racist KKK members shouting "white power", whilst encountering a crowd of black protesters shouting "black power". Intergroup contact, empathy, education here failed. The members of the extremists groups were very fixed about their ideas, and for some it might be difficult to change their views. Also the question here is again the only motive is racial prejudice, or is it also belongingness, fear of limited resources, etc. In any case, should in such cases prejudice be reduced – pharmacologically – if that was possible?

Kelly and Roeder (2008) discussed in their philosophy paper two questions; (a) If it is morally condemnable to harbour implicit biases and (b) if people's implicit biases should be corrected. The authors suggest that the answer to both questions were complex but that there are many philosophical arguments leading to answering them both with "yes". However, they also note that as people might not be aware of their implicit biases that:"… one might say that such attitudes are morally wrong – and condemnable – but that the person himself cannot be blamed for having them." However, something apparently morally condemnable, such as aggression and violence might also be needed at times. First, consider the case of your own child. Most people would agree that providing additional benefit or help towards your own child is morally acceptable. But strictly speaking, this would be prejudice. Furthermore, consider the case of brutal murders or rapists. Most people would agree that we would need some form of punishment for such people, partly through abstract considerations of justice and deterrence, but partly through emotional reactions of anger. Anger here seems to have positive function, as it can be used to punish antisocial individuals and make society work. Thus, prejudices against gross anti-social behaviour, and prejudice in terms of favouring one's own child, most people would usually not be considered as "bad". Indeed, traits that superficially might seem "negative" can also be beneficial in certain situations. For instance Ronson (2011) in his book "The Psychopathy Test" reported evidence that some "psychopaths" may be disproportionately represented in top work positions. Many might have heard about the ruthless, less empathic "business man" idea. However, a lower degree of empathy might also be useful in some types of situations not seemingly related to psychopathy. One would not want their surgeons to have too much empathy and thus not operate. However, more important is the fact that no one knows what the future holds and therefore, attributes that might be mostly a disadvantage was now – in our society – might bring a benefit later. For instance imagine aliens invade, and they are aggressive and want to kill us. With no prejudice we would hopelessly die out. The point here is that reducing variability of traits (characteristics) might seem a good idea for now, but might make humans less flexible in the future. So if it is mostly seen as "good" to favour your own child over a stranger, what then is equality?

6.2   What Is Equality? What Do We Want?

"I have a dream that one day this nation will rise up and live out the true meaning of its creed: We hold theses truth to be self-evident: that all men are created equal." (Martin Luther King, Jr.)

What is equality? Or what should equality be? Parfit (1997) described the philosophical perspective of Nagel; "imagine that you have two children, one of them healthy and one of them suffering from a painful disability. You could now move with your family to the city, which would allow the disabled child to receive special treatment. However, you could also move with your family to the countryside where the healthy child would flourish. Furthermore, you know that the benefit of moving to the countryside has a substantially greater gain to the healthy child then the gain the disabled child would have if you were to move to the city." Nagel argues that helping the disabled child is an egalitarian decision, as you are creating equality, even though the benefit you give is less than the one you could give to the healthy child. From a utilitarian perspective however, it would be considered 'good' to create the most positive benefit overall, which would be to help the healthy child. So what is fair? Does everyone have to have the same? Are people worse off in morally significant ways, and – in the extreme – do you have to give a blind stranger one of your eyes? Can you give yourself more than others? Or your family?

Kelly and Roeder (2008) discussed affirmative action; a philosophical and political term for activities in which minority groups receive benefits beyond what is warranted by the merits of the individuals in order to promote higher order moral or political motives such as complete equality. Anderson (1999) stated that: "The proper negative aim of egalitarian justice is not to eliminate the impact of brute luck from human affairs, but to end oppression which by definition is socially opposed, …, to create a community in which people stand in relations of equality to others." It might be considered a sensitive statement to suggest that there are *no* differences between nationalities, between genders, ages etc. This might lead back to the debate of whether anything that involved stereotypical ideas could be termed as racist. In fact if one argues that (a) we are all the same and (b) we should come together to increase diversity, then this seems circular. Steven Pinker wrote, referring to Jussim (1995) that:" Politically correct sensibilities may brindle at the suggestion that a group of people, like a variety of fruit, may have features in common, but if they didn't there would be no cultural diversity to celebrate and no ethnic qualities to be proud of. Groups of people cohere because they really do share traits, albeit statistically. So a mind that generalises about people from their category membership is not ipso facto defective. … business students *are* really more politically conservative than students in the arts – on average". It could be considered therefore that the statement "we are all the same", should be modified to "we all should be treated the same". In 1792 equality and rights – in this case women's right – were debated. A milestone in history was Wollstonecraft publication of an article entitled "A Vindication of the Rights of Woman", in which she argued that women should be treated equally to men. She did not make the argument for gender equality though on the basis of stating that women and men *are* equal but rather on the noting that women and men are equal in the eyes of god (e.g., in a moral sense), and that thus the same moral laws should apply to them. Furthermore, it can be argued that everyone is equal as an individual but every individual is different.

In this final chapter, I also decided to insert a section on individuality, because it is indeed essential to the understanding of the concepts. Let me start with the question; what makes a person anti-social. An uncountable number of research studies have been conducted, trying to address this question. We find that genetic factors play a role. We find that certain gene combinations might partly pre-determine certain behaviours. We find that drug abuse during pregnancy can also have an effect on the development of the child's later behaviour. Birth difficulties can have long-term effects. The parent's relationship to the child and their upbringing can contribute to the development of anti-social behaviour. Furthermore, friends, peers, and teachers can contribute to the child's behaviour. The external support the child has might play a role. The personality of the child might play a role; but this might interact with the behaviour of the parents. Concentrations of certain neurotransmitters have been found to correlate with anti-social behaviour tendencies. Gender might play a role, and the time of the day plays a role. The accessibility of weapons plays a role. Media and violent video games can contribute. The society and country in which the person lives also plays a role. Yes, all that. I believe it has become clear that studies have found a large number of factors that are relevant and I have not even listed all of them. Thus, it is clear that in order to predict if someone is likely to be or become anti-social is a complex and near impossible task. And now you want to predict someone's behaviour merely by one factor, such as their gender or ethnicity? Good luck. This is the first idea that I wanted to illustrate in this book; why prejudice is wrong. I believe that it is not only morally wrong but also because it is highly inaccurate to predict a person by one factor, even if that factor might have a kernel of truth. I wanted to suggest that all humans are individuals and all different in their own way. Furthermore, I also wanted to illustrate another factor with regards to individuality, which complicates research and results in finding that they are less clear then it might seem at first glance. Any research in psychology always relates to average effects, which means that we find results, such as playing violent video games increases aggression, but this only holds on average. This means that in no single study does someone find a 100 % result. For instance in our propranolol group, the drug reduced racial biases, but only on average, so not in every single participant. Indeed, it might even be that it increased racial biases in one person, but on average biases were lower for that group. We only always find average effects because humans are all very different and their responses to certain manipulations vary a lot. This also means that there is no effect, or treatment, or indeed any drug, which produces the same and one effect in literally everyone.

Thus, it is much too simple to presume that there was one drug that did one thing in everyone, and now all we need to do is to take it. Humans are individuals and too complex to support such an idea.

In 1948 the endorsement of the Universal Declarations of Human Rights by forty-eight countries signalled a major event. The declaration states that:

Article 1: All human beings are born free and equal in dignity and rights. They are endowed with reason and conscience and should act towards one other in a spirit of brotherhood.

Article 2: Everyone is entitled to all rights and freedoms set forth in this declaration, without distinction of any kind, such as race, colour, sex, religion, language, political or other opinion, national or social origin, property, birth or status. Furthermore, no distinction shall be made on the basis of the political, jurisdictional or institutional status of the country or territory to which a person belongs, whether it is independent trust, non-self-governing or under any other limitations of sovereignty.

Article 3: Everyone has the right to life, liberty, and security of person.

Indeed, Steven Pinker discussed abstract moral concepts, in which he suggested that expending empathy might have not been the only cause for a reduction in violence, but that an expanding circle of rights led to the reduction in violence. Specifically there is an increasing acceptance that all human beings have the same moral rights, then – even if you don't love everyone, or feel empathy – we have the knowledge that everyone deserves to be treated equally. In his complex philosophical book Lippert-Rasmussen (2014) discussed discrimination and why it is wrong based on philosophical accounts of harm, of meaning, and of mental state. The author defined generic discrimination with a complex formula (i.e., starting with an agent, X, discriminating against someone, Y, in relation to another, Z, by Ω-ing (e.g., hiring Z rather than Y). He explains that in this basic sense discrimination involves treating someone disadvantageously to others because he or she has or is believed to have some particular feature that those others do not have. Furthermore, as discussed previously, individuals seem much too complex to assess their behaviour on the basis of *one* single factor. Everyone is an individual, with moral equality and rights. Thus, treating everyone as a moral equal seems to be the aim. Therefore, favouring oneself, or one's own child, and thus showing some "prejudice" does not prevent someone from avoiding discrimination and treating everyone as a moral equal. And indeed, affirmative action (e.g., giving the minority group more) might also be seen as treatment by categorical membership. Treating everyone as a moral equal and seeing everyone as an individual is thus the moral enhancement that could be achieved partly independently of "curing prejudice". Furthermore, as discussed in the beginning, even though prejudice might be negative today, reducing genetic variability and eliminating any form of biases, might make humanity vulnerable, as we cannot predict the future, when what is currently considered to be undesirable could be necessary for survival. Thus, finally; "Prejudice can we cure it?" the answer is; no. The seemingly ordinary man at the beginning of the book, who turned out to be a Nazi, killing innocent people in a concentration camp, was probably also behaving badly because of social and political forces. He probably would not have taken a drug anyway. Furthermore, the book should have illustrated that even though we understand that everything is associated with activity in the brain, it does not mean that only a drug can change it. The IAT for example suggests that we have implicit biases, but can decide how to behave when considering moral equality. It is indeed startling that we can investigate and modify implicit biases experimentally, which gives us a greater understanding of brain function, even though an individual's level of prejudice might be stronger influenced by social manipulations.

Chapter 6 Open Questions
- What do you think is equality?
- Do you think prejudice can ever be good?
- What is moral enhancement (i.e., what is the end state that we want to achieve?)?
- Would you value diversity over limitations?

References

Anderson, E. S. (1999). What is the point of equality? *Ethics, 109*, 287–337.

Blum, L. (2004). Stereotypes and stereotyping: A moral analysis. *Philosophical Papers, 33*(3), 251–289.

de Waal, F. (2010). *The age of empathy: Lesson for a kinder society*. London: Souvenir Press.

Gilbert, P. (2010). *The compassionate mind*. London: Constable & Robinson.

Greene, J. (2013). *Moral Tribes*. New York: Penguin Press.

Haidt, J. (2007). The new synthesis in moral psychology. *Science, 316*, 998–1002.

Hartup, W. W. (1979). The social worlds of childhood. *American Psychologist, 34*(10), 944.

Hobbes, T. (1651). *Leviathan*. New York: Oxford University Press.

Jussim, L. J., McCauley, C. R., & Lee, Y. T. (1995). *Stereotype accuracy*. Washington, DC: APA.

Kelly, D., & Roeder, E. (2008). Racial cognition and the ethics of implicit biases. *Philosophy Compass, 3*, 522–540.

Krzanaric, R. (2014). *A handbook for revolution; Empathy*. London: Random House.

Lippert-Rasmussen, K. (2014). *Born free and equal*. Oxford: Oxford University Press.

Parfit, D. (1997). *Equality and priority*. Oxford: Blackwell.

Pinker, S. (2011). *Better Angels of our nature*. New York: Penguin Press.

Poussaint, A. F. (1999). They hate, they kill, they are insane? *New York Times*, 7 Dec 2008.

Ronson, J. (2011). *The psychopathy test*. London: Pan Macmillan.

Singer, P. (1981). *The expanding circle: Ethics and socio-biology*. New York: Princeton University Press.

Widiger, T. A. (Ed.) (2012). *The Oxford handbook of personality disorders*. Kentucky: Department of Psychology.

Zeitfracht Medien GmbH
Ferdinand-Jühlke-Straße 7
99095 Erfurt, Deutschland
produktsicherheit@kolibri360.de